KT-151-448

CHEMISTRY
at work
comprehension exercises
for advanced level

J. R. TAYLOR
MSc, PhD, CChem, FRSC

Head of Science
King's College
Madrid

JOHN MURRAY

To Genevieve

© J. R. Taylor 1992
First published 1992 by
John Murray (Publishers Ltd)
50 Albemarle Street, London W1X 4BD

All rights reserved. Unauthorised
duplication contravenes applicable
laws

A CIP catalogue record for this book is available from the
British Library

ISBN 0 7195 5039 4

Set by Wearside Tradespools, Boldon, Tyne and Wear, in
10 on 12pt Century Schoolbook
Printed in Great Britain by Biddles Limited, Guildford

Contents

Introduction

These exercises are designed to broaden and deepen 'A' and 'AS' level students' experience of chemistry. They begin with a passage adapted from a recently published book or journal, followed by 6–9 questions designed to test the readers' understanding of the text itself and of the chemical concepts therein. Whilst a few of the passages could be described as 'chemists' chemistry', most deal with the involvement of chemistry in industry, medicine, the environment, etc., together with the ensuing benefits and problems; in brief, Chemistry at Work. It is hoped that the passages are sufficiently relevant to stimulate interest in aspects of the subject, and that the questions will help students to gain confidence in manipulating the ideas which they have encountered in their chemistry course. In addition, an increasing number of Boards include a comprehension exercise in the 'A' level examination, for which these exercises will provide valuable practice material.

I have tried to ensure that all of the typical 'A' level topics are covered, with most of the exercises ranging quite widely over the syllabus. To help teachers decide when a particular exercise is appropriate to their students, I have included at the head of every exercise a list of the topics covered in that exercise. There is also an alphabetical topic index at the end of the book.

In the questions which follow the exercises, the number in square brackets is the suggested mark allocation. Answers to numerical problems are on p. 121. These can be detached. Suggested answers to the comprehension questions may be obtained free of charge on application to John Murray Educational Department. Please quote ISBN 0 7195 5113 7.

At the end of most of the exercises there is some additional work, usually in the form of a more open-ended question, which can be used as a basis for discussion of a wider issue raised by the text, or for some library research. The teacher is, of course, free to invent his or her own questions to accompany the text.

I would like to thank the following for permission to use published material. In all cases the source has been acknowledged beneath the text.

Blackwell Scientific Publications
Cambridge University Press
The Royal Society of Chemistry, publishers of *Chemistry in Britain*
Oxford University Press
Prentice Hall International Inc.

Last, I would like to thank the Sixth Form chemists of King's College, Madrid, who have tested the material patiently, and whose answers and suggestions have led to much revision.

It is important to note that none of the passages in this book are intended as instructions for practical work. Would-be experimenters are urged to refer to the original texts.

J. R. Taylor
Madrid

MAINLY ABOUT ELEMENTS

Exercise 1·1

The anomalous nature of lithium

equations, formulae, Group I, polarisability, solubility, thermochemistry

In general, the differences between the first and second members of any main group of the Periodic Table are greater than those between other consecutive members of that group. These differences are of degree, rather than of kind, and are amply demonstrated in the case of lithium. Almost all of these 5 differences can be traced to the smaller size of the lithium atom and its cation.

It has stronger intermetallic bonding than the other Group I metals, and this results in a high enthalpy change of atomisation, hardness, melting point and boiling point. The lattice 10 energies of lithium salts are numerically large, and because of the relatively high electronegativity of the metal its compounds have appreciable covalent character. To illustrate the effects of some of these differences, a few of the reactions of lithium and its compounds are compared with those of the 15 other alkali metals.

1. The metal reacts much more slowly with water.

2. It burns in air to give the simple oxide, instead of the higher oxides formed by the other members of the group.

3. Lithium alone in Group I will react directly with elemental 20 nitrogen to form the nitride.

4. Compounds containing anions of high charge density, such as the hydroxide, fluoride, phosphate and carbonate, are less soluble in water. This is perhaps surprising in view of the very exothermic hydration energy of the cation. 25

5. Salts with large, polarisable anions are much less thermally stable. The carbonate will decompose at 700 °C, whereas that of sodium is stable to a much higher temperature. Like most nitrates, but unlike those of the other alkali metals,

lithium nitrate loses nitrogen(IV) oxide at the temperature 30
of a bunsen flame. The hydroxide too is relatively unstable,
and it dehydrates readily when heated.

6. The halides of lithium are more soluble in organic solvents.

7. The other alkali metals will form double sulphates with
   triply-charged cations (the so-called alums), such as potas- 35
   sium aluminium sulphate dodecahydrate. The lithium
   cation, however, is too small to form a stable lattice in these
   circumstances.

Due to the trends across periods and down groups, we often
find diagonal relationships in the Periodic Table in which the 40
first member of a group shows remarkable similarities to the
second member of the group to the right of it. This is exem-
plified by lithium and magnesium; the latter metal also forms
a nitride by direct combination, and its compounds are generally less soluble and less stable than the compounds of the rest 45
of the Group II elements.

1 Why are the atoms of the first member of a group the smallest? [2]

2 Write the formulae of
   a the simple oxide of lithium (line 18), and
   b potassium aluminium sulphate dodecahydrate (lines 35–6). [3]

3 Write balanced equations for:
   a the reactions of lithium and magnesium with nitrogen (lines 20–
   21, 43–4);
   b the thermal decomposition of lithium nitrate (line 29). [6]

4 Explain why the intermetallic bonding in lithium is stronger than in
   the other Group I metals (line 8). [2]

5 What evidence given in the passage would explain why the halides
   of lithium are less water soluble (line 33)? [2]

6 Briefly discuss the reactivity trend going down Group I. What
   factors effect this trend? [5]

7 Define the term 'hydration energy' (line 25). Why is it surprising that
   a highly exothermic hydration energy should accompany a low
   solubility, and what other energy consideration might explain the
   apparent paradox? [5]

8 What is meant by 'polarisable anions', and why should such anions be less thermally stable when combined with lithium than when combined with the other elements in Group I? [5]

Total = 30

Additional work

Find out about some of the uses of lithium and its compounds.

Solvated electrons

activation energy, electrolysis, Group I, solvation

Liquid ammonia, which may be formed by cooling the gas to
below $-33\,°C$, is an example of a non-aqueous solvent which
will dissolve ionic species. Many analogies may be drawn
between reactions which occur in ammonia and those in water.
Both solvents are slightly self-ionising. In water we define an   5
alkali as a substance which dissolves giving OH^- ions; so in
liquid ammonia we may define an alkali as a substance which
dissolves to give NH_2^- ions. Thus $NaNH_2$ is an alkali in
ammonia (equivalent in water: NaOH), and so is K_2NH
(equivalent in water: K_2O). Ammonium salts may be regarded   10
as acids in liquid ammonia, and a typical neutralisation
reaction might be:

$$NH_4Cl + KNH_2 \rightarrow KCl + 2NH_3$$

One of the most remarkable properties of the alkali metals is
their ability to dissolve in liquid ammonia, forming solutions   15
which in some ways resemble liquid metals. On evaporation of
the solvent, the solid metal is reformed (except in the case of
lithium). It is known that these solutions contain alkali metal
cations and electrons, e.g.:

$$Na_{(s)} + am \rightarrow Na^+_{(am)} + e^-_{(am)}$$ 20
$$(am = liquid\ ammonia)$$

The removal of an electron from an alkali metal in the gas
phase is highly endothermic, but in solution both the metal
cation and the electron are heavily solvated, i.e., surrounded
by ammonia molecules which are tightly held by ion-dipole   25
forces, so the overall reaction is exothermic. The properties of
the solution are determined largely by the properties of the
solvated electron, which we can think of as the simplest
odd-electron species. The electron is highly mobile, and so the
solution has a high conductivity, higher than solutions of most   30
alkali metal salts. Alkali metal ions are colourless, but the
solvated electron absorbs in the red and infra-red region of the
electromagnetic spectrum. Thus the solutions of alkali metals
in ammonia all have the same blue colour.

At first sight it might seem that there are few similarities   35
between the actions of water and ammonia on the alkali
metals. Ammonia produces a solution containing electrons,

whereas water reacts readily evolving hydrogen. The reason for the difference is the variation in the stability of the solvated electron in the two solvents. It has been shown that the 40 lifetime of an electron in water is less than one thousandth of a second. It reacts as follows:

$$e^-_{(aq)} + H_2O_{(l)} \rightarrow OH^-_{(aq)} + \tfrac{1}{2}H_{2(g)}$$

We may therefore imagine the reaction of sodium with water as involving the formation of Na^+ ions and electrons, followed 45 by reaction of the electrons with the solvent. In contrast, the electron is stable almost indefinitely in ammonia if the solvent is pure, and so a metal solution is formed. If a suitable catalyst such as iron(III) oxide is added, however, then decomposition is rapid: 50

$$e^-_{(am)} + NH_{3(l)} \rightarrow NH^-_{2(am)} + \tfrac{1}{2}H_{2(g)}$$

which is precisely analogous to the reaction in water. The reaction of ammonia with the electron is energetically favourable, but in the absence of a suitable catalyst the activation energy is too high for the reaction to occur at a suitable rate. 55 The catalyst reduces the activation energy by providing an alternative pathway, and decomposition then occurs rapidly.

Adapted from Odd Electron Species, *P. R. Scott,*
Cambridge University Press (1981)

1 Write equations for the self-ionisation of water and ammonia (line 5). [4]

2 Write an equation for the reaction of K_2NH with liquid ammonia (line 9). How is this analogous to the corresponding aqueous reaction? [4]

3 What name is given to the removal of an electron in the gas phase (lines 22–3)? Give a *precise* definition, illustrating your answer with an equation for sodium. [4]

4 Explain the meaning of the terms
  a odd-electron species (line 29);
  b catalyst (line 48);
  c activation energy (line 56). [6]

5 Draw diagrams to illustrate the way in which a sodium cation and an electron become solvated, using *four* solvent molecules for each, clearly indicating the method of bonding (lines 23–6).   [4]

6 Draw a suitable energy level diagram for the processes which occur during the dissolution of sodium in liquid ammonia.   [3]

7 What reactions, if any, will occur at the anode and cathode during the passage of an electric current through a solution of sodium in liquid ammonia? Is the author justified in referring to 'conduction' rather than 'electrolysis' (lines 29–31)?   [5]

Total = 30

Exercise 1·3

Giant fullerenes—nature's footballs

bonding, carbon, hybridisation, mass spectrometry,
structure

In September 1985 chemists at the University of Sussex laser-vaporised graphite into a helium atmosphere, and the carbon species formed were analysed by mass spectrometry. The mass spectrum showed that a stable C_{60} molecule was formed. The experiments had originally been initiated to probe 5 the theory that the long carbon chain molecules—such as HC_5N—found in space might have been formed under the high temperature and pressure conditions that prevail in the atmospheres around stars (the 'circumstellar' atmosphere). The experiments readily confirmed that long chains with as many 10 as 20–24 carbon atoms could form in circumstellar atmospheres. The amazing C_{60} result stimulated a search for a structural explanation.

It did not take long before a possibility suggested itself; the stability might be the result of the hexagonal sheet curving to 15 form a 'chicken wire' cage. Such a structure would eliminate the dangling bonds of a flat graphitic sheet, and make a highly resilient cage. This would resemble the fascinating Geodesic Dome which Buckminster Fuller had designed for the US pavilion at Expo 67 in Montreal. Hence the name 'fullerenes' 20 was given to this type of compound containing carbon rings in closed cages. The giant C_{60} species, structure (1) in Figure 1, is actually called buckminster-fullerene!

The possible existence of closed-cage carbon molecules was first suggested in 1966 by David Jones, writing under the 25 pseudonym of Daedalus, in the *New Scientist*. Several have now been detected in the products of experiments like the one outlined above. It has also been suggested that these structures will always result when the gas phase nucleation of carbon occurs, such as in the formation of soot from a candle 30 flame.

Many fullerenes have been modelled, either by actual molecular models (1) or by computer simulations, up to several thousand of C atoms, e.g., (2) and (3). Of all the fullerenes, the C_{60} species is the most stable. The smallest stable species has 35 28 carbon atoms (4), and has the formula $C_{28}H_4$. As a result of these studies, calculations and experiments, some general points can be made.

(1) C_{60}

(2) C_{240}

(3) C_{540}

(4) $C_{28}H_4$

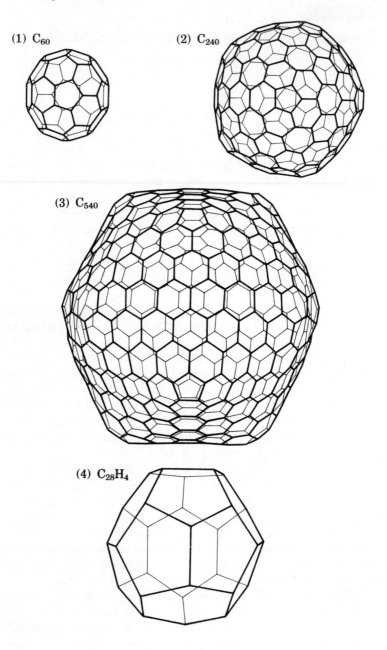

Figure 1 Structures of some giant fullerenes.

1. The smaller fullerenes are spherical, but the larger ones have more complex shapes, as shown by computer simulations. All must, however, have an even number of C atoms.   40

2. It is not possible to produce a curved surface made entirely of six-membered rings, so all the fullerenes will contain some five-membered rings as well—rather like the surface of a football!   45

3. Although in most fullerenes the C atoms all have sp^2 hybridisation, and the fourth electron is delocalised, some will also contain a small number of sp^3 hybridised carbons, and hence hydrogen is needed for the fourth valence electron; for example, the compound shown in structure (4).   50

Adapted from an article by H. Kroto in Chemistry in Britain, *January 1990*

1 Why do you think that a helium atmosphere was used in these experiments (line 2)?   [2]

2 Draw a displayed formula of the linear molecule HC_5N (line 7).   [2]

3 In what way do you think the conditions used in these experiments might resemble those found in the circumsteller atmosphere (line 9)?   [3]

4 Examine the model of the $C_{28}H_4$ molecule (4). The four H atoms will be attached to the C atoms where three 5-membered rings meet. Copy or trace the diagram, and circle these C atoms.   [2]

5 Describe the bond breaking and making processes occurring in the conversion of candlewax to buckminster-fullerene (lines 29–31).   [5]

6 Explain carefully what is meant by the terms
 a sp^3 and sp^2 hybridisation;
 b delocalised (lines 46–8).   [8]

7 Some footballs are made of triangles and pentagons. Do you think that equivalent carbon structures might form? Justify your answer.   [3]

Total = 25

Additional work
Draw a simple, labelled, diagram and write a paragraph to explain how a mass spectrometer (line 3) works.

The allotropes of phosphorus

allotropy, bonding and structure, electronic structure,
intermolecular forces, Lewis structures, solubility

There is only one known allotrope of nitrogen: the colourless
diatomic gas. In contrast, phosphorus has several allotropes,
all of them solid. This can be attributed to the fact that the
second member of Group V cannot form strong multiple bonds
to itself, and must therefore satisfy its minimum valence   5
requirements by other means.

White phosphorus forms when the vapour condenses. It is
a translucent yellowish solid, consisting of P_4 molecules held
together by weak van der Waal's forces. Consequently it is soft,
has a low melting point and is appreciably volatile at room   10
temperature. It is soluble in several organic solvents, both
polar and non-polar, but not in water. Indeed, it is stored under
water as it is very reactive and can burst into flames in the air,
particularly when finely divided or when warmed. The vapour
in contact with air gives off a pale greenish glow, called   15
'phosphorescence', due to complex oxidation reactions.

The chemical reactivity of the white allotrope is due to the
highly strained 60° bond angles in the pyramidal molecule. On
prolonged exposure to light or on heating it polymerises by
opening one of the bonds (Figure 1), leading to a much less   20
strained form, **red phosphorus**. This is a dark violet-red,
opaque solid, much less reactive, and insoluble in either water
or organic solvents. It has, of course, a much higher melting
point, and is involatile at room temperature: for this reason it
is very much less toxic. It is the only stable allotrope of this   25
element, the other structures being metastable. That is to say,
they are energetically unstable with respect to red phosphorus,
and will change into it at a rate which depends upon temperature.

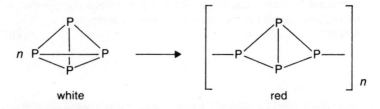

white   red

Figure 1 The polymerisation of white phosphorus.

A metallic allotrope, **black phosphorus**, is made by sub- 30
jecting the red form to very high pressure. It resembles
graphite in appearance and in properties: it has a dark grey,
metallic lustre, is flaky, and is a conductor of heat and
electricity. Not surprisingly, it has a layered structure very
like that of graphite. Each atom is bonded covalently to three 35
others, but the four form a pyramidal rather than a planar
structure. This leads to corrugated hexagonal plates weakly
bonded together (Figure 2).

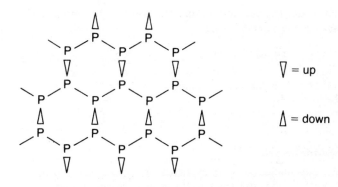

∇ = up

Δ = down

Figure 2   Black phosphorus, showing part of one layer.

A fourth allotrope, **brown phosphorus**, is only stable at
very low temperatures ($< -196\,^{\circ}$C). It is prepared by rapidly 40
condensing the vapour, and since the latter consists of P_2
molecules, it is reasonable to suppose that these are also
present in this solid form. It is the most reactive allotrope.
Other allotropes have been reported: scarlet, violet, and
another black form, but very little is known about them. It is 45
possible that they are variations of the red or black varieties.

1 What are the 'minimum valence requirements' of phosphorus
(lines 5–6)? Explain your answer with the aid of an appropriate
diagram of the electronic structure of the element.   [3]

2 Explain the origin of van der Waal's forces, with reference to white
phosporus (line 9).   [4]

3 Give one example each of a polar and a non-polar organic solvent
(line 12).   [2]

4 How would you explain the insolubility of white phosphorus in
  water (line 12)?   [2]

5 Explain how the strain is relieved by polymerisation (lines 20–
  21).   [2]

6 Why 'of course' (line 23)?   [2]

7 Would you expect black phosphorus to have the slippery feel of
  graphite? Justify your answer.   [3]

8 How might it be possible for the black allotrope to conduct heat
  and electricity (lines 33–4)?   [2]

9 Draw a Lewis structure (dot-and-cross diagram) of the probable
  electronic structure of the P_2 molecule. By considering the nature
  of the bonds in the two molecules, explain why P_2 is so much less
  stable than N_2.   [5]

                                                        Total = 25

Additional work

What is the principal ore of phosphorus, and how is the element
extracted from the ore? What are the uses of phosphorus and its
compounds?

The oxides of nitrogen

*environment, equations, equilibrium, Le Chatelier's
principle, Lewis structures, oxidation numbers,
shapes of molecules*

Nitrogen forms compounds in which it shows all oxidation
states from -3 to $+5$, and there is an oxide for each of the
positive states—i.e., it has five oxides, which is probably more
than any other element.

Nitrogen(I) oxide (nitrous oxide), N_2O

5

Commonly called laughing gas, since inhaling small amounts
can give rise to symptoms of intoxication; larger quantities
induce sleep, and it is the earliest known anaesthetic, still used
for this purpose by dentists. It is also used as an aerosol
propellent. It is prepared in the laboratory by carefully heating   10
ammonium nitrate to 200 °C:

$$NH_4NO_{3(s)} \rightarrow N_2O_{(g)} + 2H_2O_{(l)}$$

On heating, this oxide decomposes into its elements, and
hence will re-kindle a glowing splint.

Nitrogen(II) oxide (nitric oxide), NO

15

Like the previous oxide, it is a colourless, neutral gas, but
slightly toxic. It may be prepared by the action of 50% nitric
acid on copper:

$$3Cu_{(s)} + 2NO_{3(aq)}^- + 8H_{(aq)}^+ \rightarrow 3Cu_{(aq)}^{2+} + 2NO_{(g)} + 4H_2O_{(l)}$$

It is also formed by direct combination of the elements at   20
high temperature, and hence is a constituent of vehicle exhaust gases. It reacts with dioxygen to form nitrogen(IV) oxide,
and with ozone to form dioxygen:

$$O_{3(g)} + NO_{(g)} \rightarrow O_{2(g)} + NO_{2(g)}$$

Ozone occurs in the upper atmosphere, and protects the   25
surface of the Earth by absorbing harmful UV radiation, hence
the presence of NO in the atmosphere is undesirable. Furthermore, the nitrogen(IV) oxide produced reacts with atomic
oxygen (also found in the upper atmosphere) and NO is
regenerated. Hence NO is a catalyst, and a small amount can   30
destroy large quantities of ozone.

Nitrogen(III) oxide (dinitrogen trioxide), N_2O_3

This oxide, which can be regarded as the anhydride of nitric(III) acid, is very unstable and only exists as a pale blue solid at low temperatures. It can be prepared by mixing equal   35
volumes of NO and NO_2 and cooling to below $-20\,°C$. The reaction reverses on warming.

Nitrogen(IV) oxide (nitrogen dioxide), NO_2

This dark brown oxide is V-shaped, and contains a delocalised molecular orbital. Like NO, it has an unpaired electron on the   40
nitrogen atom, which enables the molecule to dimerise; the two forms are in equilibrium:

$$2NO_{2(g)} \rightleftharpoons N_2O_{4(g)} \qquad \Delta H^\ominus = -58\,kJ\,mol^{-1}$$

The position of equilibrium shifts to the left as the temperature increases and pressure decreases; dissociation is virtually   45
100% complete at 150 °C. NO_2 is prepared by heating lead(II) nitrate, which decomposes to give a mixture of this oxide and oxygen. It is toxic, and dissolves in water to give a mixture of nitric(III) and nitric(V) acids.

Nitrogen(V) oxide (dinitrogen pentoxide), N_2O_5   50

This is a colourless solid at room temperature, which can be prepared by heating nitric acid with phosphorus(V) oxide. In the gaseous state, the molecule has the formal structure

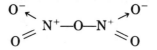

but again there is delocalisation at the O—N—O groups. In the   55
solid state it is ionic, $NO_2^+\,NO_3^-$, nitronium nitrate. The nitronium ion is isoelectronic with carbon dioxide, and the nitrate ion with the carbonate ion.

1 Explain what happens when N_2O relights a glowing splint, giving a relevant equation (line 14). [4]

2 What other product is formed when nitrogen(IV) oxide reacts with atomic oxygen (lines 28–30)? Explain how nitrogen(II) oxide catalyses the destruction of ozone. [4]

3 Write balanced equations for the reaction of nitrogen(III) oxide
   a on adding it to water, and
   b on warming the gas (lines 33–7). [4]

4 Write a balanced equation for the decomposition of lead(II) nitrate
   (lines 46–7). [2]

5 Draw a displayed formula of NO_2, clearly showing the unpaired
   electron and the delocalised system. Use this to indicate how the
   molecule dimerises (lines 39–41). [4]

6 Explain, in terms of Le Chatelier's principle, the effect of
   temperature and pressure on the equilibrium referred to in
   line 43. [4]

7 What is meant by 'isoelectronic' (lines 57–8)? Draw Lewis
   structures (dot-and-cross diagrams) of the nitrate and carbonate
   ions to illustrate your answer. What shape do these two ions
   have? [6]

8 What risks might be attached to the use of N_2O as an aerosol
   propellent (lines 9–10)? [2]

Total = 30

Additional work

In view of the fact that vast amounts of nitrogen(II) oxide are produced
naturally by electrical storms, is the costly process of eliminating this gas
from vehicle exhausts really worth it?

Interhalogens and polyhalogens

bonding, diffusion, Lewis structures, redox equations,
shapes of molecules

There are many compounds and ions with halogen-halogen bonds. They are of special importance as highly reactive intermediates, and for providing useful insights into bonding. Binary interhalogens have the formulae XY, XY_3, XY_5, and XY_7, where X is the heavier, less electronegative halogen.   5

All possible diatomic molecules XY are known, but many are thermally unstable. The most stable is ClF, but ICl and IBr can also be obtained in pure, crystalline form. Their physical properties are intermediate between X_2 and Y_2; for example, we may compare the deep red ICl (mp 14 °C, bp 97 °C) with   10
green Cl_2 (mp -101 °C, bp -35 °C) and black I_2 (mp 114 °C, bp 184 °C). Interestingly, the physical properties of ICl, prepared by simply passing chlorine over iodine at room temperature, are very like those of Br_2.

Most of the higher interhalogens are fluorides. The only   15
neutral compound in which the central atom has an oxidation state of $+7$ is IF_7, but ClF_6^+ is known. The shapes of all these molecules are as predicted by theory. For example, the XY_3 compounds (such as ClF_3) are slightly distorted T-shapes, as a result of 5 pairs of electrons in a trigonal bipyramidal arrange-   20
ment (structure (1) below), in which the F—Cl—F bond angle is a little less than 90°. However, ICl_3 is a bridged dimer (2). XY_5 compounds are almost square-based pyramids, again with the bond angle slightly less than 90°.

All the interhalogens are strong oxidising agents, and those   25
containing fluorine are convenient fluorinating reagents. Surprisingly, in view of the weakness of the F—F bond, they often fluorinate substances faster than elemental fluorine. Chlorine trifluoride is used to oxidise UF_4 to UF_6, the latter being a

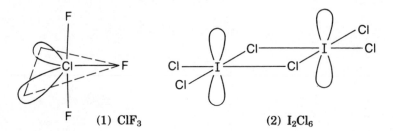

(1) ClF_3 (2) I_2Cl_6

volatile compound important for the enrichment of ^{235}U by a 30
diffusion process. Chlorine and bromine trifluorides react
violently, often explosively, with organic compounds, and will
burn asbestos. They will often expel oxygen from metal oxides;
for example Co_3O_4 reacts with chlorine trifluoride in a molar
ratio 1:3, respectively, to form cobalt trifluoride, dichlorine, 35
and dioxygen.

Bromine trifluoride is a useful solvent for reactions which
must be carried out in strongly oxidising conditions. It auto-
ionises in the liquid state, and it will dissolve fluoride salts. For
example: 40

$$CsF_{(s)} + BrF_{3(l)} \rightarrow CsBrF_{4(soln)}$$

The anion is square planar.

Many polyhalogen ions, cations as well as anions, are known.
For example, iodine will dissolve in strongly oxidising solvents,
such as fuming sulphuric acid, to form the blue diiodinium 45
cation, I_2^+; the bromine analogue can also be formed. Polyha-
lide formation is most pronounced for iodine; the best known
polyiodide is the brown I_3^- anion, formed by dissolving iodine in
a solution containing iodide ions. The bonding is very weak,
and in solution it behaves as if it were I_2; indeed, it is usually 50
referred to as a solution of iodine. It cannot be isolated in the
solid state.

Adapted from Inorganic Chemistry, *D. F. Shriver, P. W. Atkins and*
C. H. Langford, Oxford (1990)

1 Explain why the bond angles in XY_3 and XY_5 are less than 90°
(lines 21–4). [4]

2 Draw the following species, showing clearly the lone pairs:
  a IF_5 (line 23);
  b BrF_4^- (line 42).
  Give a Lewis structure (dot-and-cross diagram) for **b**. [5]

3 Show how the evidence verifies the statement that the properties
of ICl are intermediate between those of Cl_2 and Br_2 (lines 10–11).
Sketch and label an apparatus that you could use to prepare dry
chlorine and then a sample of ICl (line 13). [7]

4 What shape is the dimer shown in structure (2), and what type of
bond is responsible for the bridging? [2]

5 Explain clearly, with a suitable diagram, why the F—F bond
(line 27) is weak. [3]

6 Write an equilibrium for the autoionisation of bromine trifluoride
(lines 38–9). What shape would you predict for the cation? Draw a
suitable diagram to justify your answer. [4]

7 Write a balanced equation for the reaction mentioned in lines
33–6. Tricobalt tetraoxide is a mixed oxide; use the equation to
find the oxidation numbers of cobalt in this compound, showing
your method, and hence rewrite the formula of the oxide. [5]

Total = 30

Additional work

Outline briefly the principles of the uranium enrichment process using
UF_6 (lines 28–31).

Exercise 1.7

Some chemistry of xenon

*disproportionation, Lewis structures, oxidation
numbers, redox equations*

Until 1962 the noble gases were known as the inert gases—no
one thought it possible that they could form compounds. In
that year, however, the compound $O_2^+[PtF_6]^-$ was prepared.
The production of O_2^+ by simply mixing oxygen with an
equimolar quantity of platinum hexafluoride at room tempera-   5
ture showed PtF_6 to be 'an oxidiser of unprecedented power',
capable of oxidising molecular oxygen. Because the first ionisation energy of molecular oxygen ($1220 \, kJ \, mol^{-1}$) is slightly
more than that of xenon ($1213 \, kJ \, mol^{-1}$), it was reasoned that
xenon might be similarly oxidised. Within a month the suspi-   10
cion was confirmed—xenon was rapidly and spontaneously
oxidised by the deep red PtF_6 to give a yellow-orange solid,
$Xe^+[PtF_6]^-$.

The news of this astonishing discovery sparked off a flood of
activity, and within six months the first binary noble gas   15
fluoride, XeF_4, had been reported, followed rapidly by XeF_2 and
XeF_6. Within a year compounds of krypton and radon had also
been prepared, but to this day stable compounds of helium,
neon and argon are not known.

Xenon chemistry is the most extensive—Xe is known in   20
oxidation states ranging from $+1$ to $+8$, while krypton and
radon are known only in the $+2$ oxidation state. In a majority
of the compounds obtained, the noble gas is bonded to the most
electronegative elements fluorine or oxygen, but bonding to
chlorine, nitrogen, and carbon has also been reported. The   25
xenon fluorides are all prepared by the direct reaction between
the two gases, the product depending upon the proportions of
the reactants and other conditions such as pressure, temperature, and/or the presence of UV radiation. Apart from the
unstable free radical XeF, only even-numbered fluorides have   30
been prepared.

XeF_6 has a surprisingly high boiling point (higher than UF_6,
in fact), suggesting that it is polymeric in the solid state, but its
structure is uncertain as it hydrolyses very easily to the
explosive XeO_3. In the liquid or gaseous state it dissociates to   35
$[XeF_5]^+F^-$. Reaction of XeF_6 with alkali solution produces
'xenate' ions, $[HXeO_4]^-$. These readily disproportionate:

$$2[HXeO_4]^-_{(aq)} + 2OH^-_{(aq)} \rightarrow [XeO_6]^{4-}_{(aq)} + Xe_{(g)} + O_{2(g)} + 2H_2O_{(l)}$$

The $[XeO_6]^{4-}$ ion can be isolated as solid salts, and is amongst the most powerful oxidising agents known. It will, for example, rapidly oxidise manganate(II) ions in acidic solution to manganate(VII) ions, with the evolution of oxygen. 40

XeF_2 is stable in acidic or neutral solution, but decomposes immediately in alkali to give xenon, oxygen, and the fluoride ion. The difluoride is a mild fluorinating agent—it will react smoothly with many organic compounds, replacing H by F; for example: 45

$$CH_3I + 3XeF_2 \rightarrow CF_3I + 3HF + 3Xe$$

The cleanness (lack of side reactions) of these fluorinations makes the difluoride useful in preparing medically important fluorinated organic compounds, such as some anti-tumour agents. 50

Adapted from 25 Years of Noble Gas Chemistry, *John H. Holloway,*
Chemistry in Britain, *July 1987*

1 What is meant by the word 'oxidiser' here (line 6)? What is the oxidation number of the named element in the given formulae:
  a platinum (line 3),
  b xenon (line 13), and
  c xenon (beginning of line 36)? [5]

2 Write an equation for the first ionisation energy of the oxygen molecule, and draw a Lewis structure (dot-and-cross diagram) of the cation so formed (lines 7–9). [4]

3 Why do you suppose that no compounds of helium, neon and argon (lines 18–19) are known? [2]

4 How would altering the proportions of xenon and fluorine affect the formula of the product? What is the role of UV radiation in the reaction between these two elements (lines 27–9)? [4]

5 What is meant by a 'free radical'? Draw a Lewis structure of XeF (line 30) to illustrate your answer. [4]

6 Write balanced equations for the reaction of xenon hexafluoride with:
  a water (lines 34–5), and
  b an alkaline solution (lines 36–7). [4]

7 Write a balanced equation for the decomposition of xenon difluoride in alkaline solution (lines 43–5). [2]

8 What is meant by 'disproportionate' (line 37)? Show how this concept applies to the equation given on line 38. [3]

9 Can you suggest why the lack of side reactions when xenon difluoride is used as a fluorinating agent makes it particularly valuable in medical applications (lines 49–52)? [2]

Total = 30

The manufacture of titanium

colour in d-block compounds, electrochemistry,
electronic structure, equilibrium, industry, hydrolysis,
oxidation states, resources

Titanium ($Z = 22$), the second element in the first row of the d-block metals, was first isolated in impure form in 1826, but it was not for another 100 years that pure titanium was made by reacting the impure metal with iodine to make TiI_4, and then decomposing this on a tungsten filament at 1300–1500 °C. It   5
was found to have a high mp (1668 °C), a relatively low density ($4·51 \, g \, cm^{-3}$), and to be soft and ductile, but it can be made much harder and stronger by alloying. This method was far too expensive for the commercial production of the metal, and it was a further 15 years before an economically viable process   10
was available.

The main ores are rutile (TiO_2) and ilmenite ($FeTiO_3$): the former is used for the manufacture of the metal, as it is easier to purify. The first stage, as in the extraction of most metals, is the concentration of the ore by various standard procedures.   15
Titanium(IV) oxide, however, cannot be reduced with carbon, carbon monoxide, or hydrogen as the metal is too reactive, and even the most reactive metals will not reduce it efficiently. Cathodic reduction cannot be used either, since rutile melts at too high a temperature. The next step, therefore, is to convert   20
the white oxide to the tetrachloride. The equilibrium

$$TiO_2 + 2Cl_2 \rightleftharpoons TiCl_4 + O_2$$

lies very far to the left, however, and so rutile is reacted with chlorine in the presence of carbon at 800–1000 °C. This produces the tetrachloride and carbon dioxide. After purification,   25
the liquid titanium tetrachloride (bp = 136 °C) is stored in steel tanks under dry argon, since it hydrolyses very rapidly producing a dense white, acidic smoke.

Titanium tetrachloride cannot be electrolysed, and so it must be reduced chemically; two methods are currently in use. In   30
the first, the liquid is added carefully to magnesium at high temperature and under an inert atmosphere, when the very exothermic reduction occurs. The temperature is above the mp of magnesium chloride, but below that of titanium, and so the two are separated by running off the liquid. Both Mg and Cl_2 35
are recovered by electrolysis of this melt, and recycled. The Mg

must be used in excess in the reduction, to prevent the following reactions:

$$3TiCl_4 + Ti \rightarrow 4TiCl_3 \text{ and } 2TiCl_3 + Ti \rightarrow 3TiCl_2$$

The alternative process is almost the same, but uses sodium instead of magnesium. The tetrachloride is first reduced to $TiCl_2$ or $TiCl$ by the Na, and reduction to Ti occurs more slowly. It is not economically important to electrolyse the NaCl, so the reaction mixture is cooled, and the salt is separated from the titanium by dissolving in water.   45
Finally, the metal is purified. This is not easy, but is essentially accomplished by electric arc melting of the crude, spongy product and cooling to give blocks, which are then remelted in the same way under vacuum to vaporise impurities.   50
This necessarily complex extraction procedure means that titanium is expensive, but its unique properties make it worth it. Although it is a reactive metal, it is protected from further attack by a thin surface-coating of the oxide, so it behaves as if it were inert. Its low density, good mechanical properties   55
(particularly in alloys) and high mp make it ideal for various aeronautical engineering applications. Its chemical inertness and high mp make it suitable for electrode use in some industrial electrolytic processes, and the fact that it is not rejected by the human immune system makes it useful in   60
surgical repair work.

Adapted from The Story of Titanium, *D. J. Jones*, Chemistry in Britain,
November 1988

1 What other metal is protected by a similar thin surface- coating of the oxide (lines 53–5)?   [1]

2 By considering structures, explain briefly why alloys are harder and stronger than pure metals (lines 7–8).   [4]

3 List *all* the oxidation states of titanium given in this passage, with one example of each taken from the text.
Write the electronic structure of titanium, using the 'electrons in boxes' notation (line 1), and hence explain the existence of the various oxidation states.   [9]

4 Examine the equilibrium given on line 22, and explain why adding carbon to the reaction mixture makes it possible to make the tetrachloride in high yield (lines 23–5). Write a balanced equation for the reaction. [5]

5 What *theoretical* reason prevents the electrolysis (line 29) of titanium tetrachloride? Give evidence from the passage to justify your answer. [4]

6 Write a balanced equation for the hydrolysis of titanium tetrachloride, and use it to interpret the observed properties of this substance (lines 27–8). [4]

7 How would you classify the two reactions given on line 39? [1]

8 Why do you think that the magnesium chloride is recycled (lines 35–6), but the sodium chloride is not (line 43)? [2]

Total = 30

Additional work

Explain how colour arises in transition metal compounds, with suitable examples, and why titanium(IV) compounds are colourless. Which other metals in the first row of the transition series are colourless in particular oxidation states?

Exercise 1·9

Anticancer platinum compounds

coordination chemistry, isomerism, medicine, redox

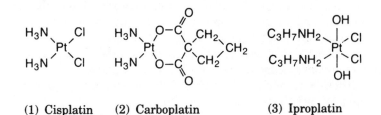

(1) Cisplatin   (2) Carboplatin   (3) Iproplatin

The discovery in 1969 of the anti-tumour activity of the square planar complex **cisplatin**, *cis*-[PtCl$_2$(NH$_3$)$_2$] (structure (1)) generated considerable interest in the pharmacology of metal complexes. The first clinical trials of cisplatin in 1971 confirmed that the compound was active against several human $_5$ tumours, but that it had side effects such as kidney toxicity, nausea and vomiting, even at low doses. These can be minimised by various stratagems, however, and the drug is currently used as part of the treatment for a variety of cancers.

Less toxic alternatives were sought, and many platinum $_{10}$ complexes were synthesised and screened for anti-tumour activity and toxicity; many *cis*-diamines looked promising. For an homologous series of *cis*-[PtCl$_2$(amine)$_2$] complexes with alicyclic amines, anti-tumour potency decreased slightly with increased ring size, but this was accompanied by a dramatic $_{15}$ decrease in toxicity. For example, the therapeutic ratio (toxic dose:effective dose) for the di-cyclopentylamine compound was 200:1, compared to 8:1 for cisplatin, but the former was found to be too insoluble in water for a viable clinical study.

Another approach has been to replace the two Cl atoms by $_{20}$ other groups. For example, chelating malonate (mal) complexes have been prepared, *cis*-[Pt(mal)(NH$_3$)$_2$]. (Malonates are derived from malonic acid, the common name for propane dioic acid.) The most promising has been **carboplatin** (structure (2)). Clinical trials of this drug showed strong indications $_{25}$ of anti-tumour activity, with no kidney toxicity, but with some reduction in the blood count of platelets, and white and red cells. It has been found more effective than cisplatin against some tumours, but less effective against others. In 1985 a product licence was applied for in the UK. $_{30}$

A range of Pt(IV) complexes was also prepared by oxidation of the Pt(II) complexes with chlorine or hydrogen peroxide, and were found to be less toxic than the Pt(II) analogues. The most soluble, and hence the most promising, was **iproplatin** (structure (3)). In animal tests this was found to have similar ³⁵ anti-tumour activity to cisplatin and minimal kidney toxicity, with reduced emetic potential though it still depressed the white blood cell count. It has been found effective in clinical trials, even for some cancers resistant to cisplatin.

Although it is estimated that 2000 analogues of cisplatin ⁴⁰ have been synthesised in the last 20 years, research continues for compounds with even higher activity, lower toxicity, and those that can be delivered orally. The use of natural products, or 'biomolecules', as ligands for platinum is being tried. In addition, complexes of other heavy metals, like rhodium and ⁴⁵ gold, have been found to have some activity and will doubtless be focuses for research in the 1990s.

Adapted from Second Generation Anticancer Platinum Compounds,
C. F. J. Barnard, M. J. Cleare and P. C. Hydes, Chemistry in Britain,
November 1986

1 Explain concisely the following terms:
   a homologous series (line 13);
   b ligand (line 44);
   c chelating (line 21);
   d metal complex (lines 3–4). [12]

2 With reference to cisplatin, explain what *cis*- means. Draw and name an isomer of this compound. [4]

3 Draw a displayed formula of malonic acid (lines 22–4). Is the type of isomerism mentioned in the previous question possible for malonic acid? Justify your answer. [5]

4 Draw a displayed formula of the malonate complex (line 22). [3]

5 Write a balanced equation for the oxidation of cisplatin with hydrogen peroxide (lines 31–2). Indicate, using oxidation numbers, which element acts as the oxidising agent. [4]

6 Explain what is meant by the therapeutic ratio (line 16), using di-cyclopentylamine as an example to illustrate your answer. Why do you think that the insolubility of this compound in water prevented its clinical trial? [4]

7 Consider the platinum compounds given in the passage. What particular structural features, common to all the compounds, might be responsible for the anti-tumour activity? [3]

Total = 35

Additional work

Is the use of animals in tests such as those described here (lines 35–8) justified? What alternative methods of testing might be possible?

The discovery of the rare earth elements

atomic structure, fractional crystallisation, ion exchange, Moseley's experiment, Periodic Table, radioactivity, resources

Most compilations of the Periodic Table have two rows of 14 elements placed underneath almost as footnotes. The second row consists almost entirely of elements which do not occur naturally, the 'actinides', so-called because they follow actinium. The top row, elements 58–71, is now best known as the 'lanthanides', but was originally (and is sometimes still) called the 'rare earths'—a misnomer on two counts, since they are neither 'rare'—some of the more common members are more abundant than lead, tin, mercury or gold, and even the most rare is more abundant than silver—nor are they 'earths'. This latter name was given to substances which were thought to be elements as they could not be broken down into anything simpler by any method available at the time. We now know them to be metal oxides, such as lime, CaO and magnesia, MgO. Today, the lanthanides and their compounds have many uses, such as catalysts, colorants in glassmaking and ceramics, permanent magnets in electronics, in the manufacture of lasers, and in medicine.

Their story begun in 1788, when an amateur mineralogist found a strange black mineral in a quarry near Ytterby in Sweden, which he called 'ytterbite'. This was subsequently found to contain, amongst other things, a new 'earth', which we now know to be the oxide of yttrium (still called a rare earth, even though it does not belong to the lanthanide series). This stimulated the search for other 'rare earths' and over the next few years, many others (usually as their oxides) were found, but it was not until 1827 that the first lanthanide element was isolated: cerium chloride, $CeCl_2$, was reduced to the metal. The quest was to be aided later in the century by the development of new techniques such as optical spectroscopy, but many of the claimed new 'rare earths' proved to be mixtures. The main problem was their great chemical similarity, and hence separating them was extremely difficult; fractional crystallisation was practically the only method available at the time. These days, ion exchange chromatographic techniques make their separation relatively simple.

Not only were there practical difficulties associated with the discovery and isolation of lanthanides, but they presented an even greater theoretical challenge: where to place them in the Periodic Table? Mendeleyev himself, shortly before his death 40 in 1906, said that this remained one of the major difficulties of his system. The principal problem was that there was no way of knowing for certain just how many lanthanides there were. This was immediately resolved in 1913 by Moseley's discovery that the atomic number of an element could be determined 45 from its X-ray spectrum. It was now possible to state with certainty that there were 15 lanthanide elements (including lanthanum itself), of which all but one had been found by this date.

The missing one was number 61, and many chemists set out 50 to find it. Two groups claimed simultaneously in 1926 to have identified it; one group (in the USA) called theirs illinium, and the other (in Italy) florentium. An argument was prevented, however, as both were mistaken. It was not until 1947 that a tiny sample of the element was isolated from the mixture of 55 isotopes produced by uranium fission. Element 61 was in fact radioactive with a short half-life and it probably does not exist naturally on Earth. It was named promethium, after the Greek god Prometheus who stole fire from heaven and was consequently punished by Zeus. This name was intended to symbol- 60 ise the dramatic way in which the element was discovered and the benefits that can be obtained from radioactivity, and also to warn of the dangers associated with its misuse.

Adapted from an article with the same title by C. H. Evans,
Chemistry in Britain, *September 1989*

1 Outline *two* ways in which cerium chloride could have been reduced to the metal, giving appropriate equations (line 28). [6]

2 Give a brief explanation of the principles of
a fractional crystallisation (line 33), and
b ion exchange (line 35). [6]

3 How might the discovery of a new element be aided by optical spectroscopy? [3]

4 Why is it that most of the actinides (lines 2–5) do not occur naturally? [2]

5 Explain in electronic terms why there should be 14 in each of these rows. What other name is given to both rows taken together (lines 1–2)? [4]

6 What explanation, in terms of electronic structure, can you offer for the abilities of the lanthanides to act as
a catalysts, and
b colorants (line 16)? [4]

7 Explain the terms
a uranium fission (line 56);
b half-life (line 57). [6]

8 Outline briefly, including a sketch diagram, how X-ray spectra (line 46) were obtained. What important principle did Moseley discover as a result of his experiments? [4]

Total = 35

Additional work

Discuss the benefits and dangers associated with the use of radioactivity (lines 62–3).

MAINLY ABOUT COMPOUNDS

The hydrogen bond

dipoles, electronegativity, intermolecular forces, Periodic Table, thermochemistry

When hydrogen forms a covalent bond to a very electronegative atom, the bond is highly polar:

$$X^{\delta-}\!=\!H^{\delta+}$$

This dipole can interact with a lone pair of electrons on another atom, :Y—, resulting in a weak intermolecular bond (weak, that is, by comparison with a covalent bond, but stronger than other intermolecular forces, such as dipole-dipole interactions or London dispersion forces, collectively known as van der Waal's forces). Typically, bond strengths for hydrogen bonds range from about 4 kJ mol^{-1} to 25 kJ mol^{-1}. The formation of a hydrogen bond may be represented thus: 5

$$X—H + :Y— \rightarrow \quad X—H \cdots Y—$$

This type of bond uniquely has hydrogen as the central atom, due to its small size and lack of inner shells of electrons to shield the nucleus. 15

If the hydrogen bond is indeed the result of an electrostatic interaction between the X—H bond dipole and an unshared electron pair on another atom, Y, then the strength of hydrogen bonding should increase as the X—H dipole increases. Thus, for the same Y, we would expect the hydrogen-bonding strength to increase in the series 20

$$=\!N—H \cdots Y— \; < \; —O—H \cdots Y— \; < \; F—H \cdots Y—$$

This is indeed true, but what property of Y is important? As we have seen, the atom Y must possess an unshared electron pair that attracts the positive end of the X—H dipole. This electron pair must not be too diffuse in space; if the electrons occupy too large a volume, the X—H dipole does not experience 25

10

a strong, directed interaction. For this reason, we find that hydrogen bonding is not very strong unless Y is one of the smaller atoms: N, O, or F, specifically. Amongst these three 30 elements, we find further that hydrogen bonding is stronger when the electron pair is not attracted too strongly to its own nuclear centre. The energy needed to remove an electron from Y is a good measure of this aspect. For example, the ionisation energy of an unshared-pair electron on nitrogen in a covalent 35 molecule is less than the corresponding value for oxygen in the reactions

$$:NH_{3(g)} \rightarrow \cdot NH^{+}_{3(g)} + e^{-} \text{ and } :OH_{2(g)} \rightarrow \cdot OH^{+}_{2(g)} + e^{-}$$

respectively. Nitrogen is thus a better donor of an electron pair to the hydrogen in X—H. For a given X—H, hydrogen-bond 40 strength increases in the order

$$X—H \cdots F— \; < \; X—H \cdots O= \; < \; X—H \cdots N\equiv$$

When X and the donor atom Y are the same, the energy of the hydrogen bonding increases in the order

$$\equiv N—H \cdots N\equiv \; < \; —O—H \cdots O= \; < \; F—H \cdots F— \qquad 45$$

When the Y atom carries a negative charge, the electron pair is able to form especially strong hydrogen bonds. The hydrogen bond in the F—H $\cdots$ F^{-} ion is among the strongest hydrogen bonds known; the reaction

$$F^{-}_{(g)} + HF_{(g)} \rightarrow FHF^{-}_{(g)} \qquad 50$$

has a $\Delta H^{\ominus}$ value of about -155 kJ mol^{-1}. We can compare this with the energy of the strong H—F *covalent* bond (568 kJ mol^{-1}). The anomalous behaviour of some compounds can be explained by the presence of hydrogen bonding: for example, the many unusual properties of water. The fact that an 55 aqueous solution of HF is a much weaker acid than those of the other hydrogen halides can partly be attributed to its strong hydrogen bonding.

In part adapted from Chemistry, the Central Science, *T. L. Brown and H. E. LeMay, Prentice Hall (1988)*

1 What is meant by the terms
  a electronegative (line 1);
  b dipole (line 4);
  c lone pair of electrons (line 4)?
  Illustrate your answer with examples taken from the passage.   [9]

2 Explain briefly how these other two types of intermolecular force arise (lines 7–8), with one example of each. [6]

3 Why should these two factors (lines 14–15) make hydrogen unique in this respect? [2]

4 Refer to a copy of the Periodic Table. What structural feature of these three atoms (line 30) results in the opposite orders of hydrogen bond strengths, depending on whether they are attached to the hydrogen (line 22), or acting as the donor (line 42)? Justify your answer. [5]

5 Briefly describe, in principle only, how these two reactions (line 38) could be achieved, and the energies (lines 34–6) measured. [3]

6 Again referring to the Periodic Table, give an example of an atom in which the electron pair would be more diffuse, and briefly explain why this should be so (lines 26–8). [3]

7 Why should the presence of a negative charge (lines 46–7) increase the strength of the hydrogen bond? [2]

8 Explain, with a relevant equation, why the strong hydrogen bonding in a solution of hydrogen fluoride contributes to its weakness as an acid (lines 55–8).

What other, more important factor also causes this weakness? [5]

Total = 35

The formula of tin iodide

bonding, empirical formula, experimental technique,
% composition, safety

Tin and iodine react together quantitatively to form one
product, whose formula may be determined by reacting
weighed amounts of iodine with excess tin, and determining
the mass of the metal remaining after the reaction. The
procedure is described below.   5

About 5 g of iodine is weighed accurately onto a watchglass,
and transferred quantitatively to a 250 cm^3 round-bottomed
flask which is clamped securely in a stand, taking care that
none adheres to the joint by using a solid-additions funnel. A
reflux condenser is fitted to the flask. 50 cm^3 of 1,1,1-   10
trichloroethane is measured out, and added to the flask via the
condenser. The iodine will dissolve partly in the solvent.

4–5 g of tinfoil is weighed accurately, and the surface is
cleaned by wiping both sides with a tissue soaked in propanone. It is then cut into strips about 5 mm wide, and each   15
strip is loosely coiled around a pencil. These are added to the
flask, again via the condenser.

The flask is clamped in a hot waterbath, and the contents
allowed to reflux gently: the rate of refluxing may be controlled
by raising and lowering the flask. The iodine dissolves com-   20
pletely to give a purple solution and, as the reaction proceeds,
the colour changes gradually to orange. Reaction is complete
when the drops of liquid dripping from the condenser no longer
have a purple colour; this will probably take approximately 30
minutes.   25

After cooling, most of the solution is carefully decanted into
another flask using a funnel, leaving all the unreacted tin
behind. The tin is washed with three small amounts of hot
1,1,1-trichloroethane, and the washings added to the rest of
the solution. A *small* amount of solid product still adhering to   30
the tin will not affect your result, but if there is a lot, it may be
scraped off gently with a spatula, and the washing procedure
repeated. Finally, the tin is removed from the flask, allowed to
dry, and weighed. The formula of tin iodide may now be
calculated.   35

The second flask is now fitted with a stillhead and condenser,
and clamped in the waterbath. About half of the solvent is
distilled off, and the residue is cooled in ice, when orange

crystals of tin iodide are deposited. These are filtered off at the pump, washed with small amounts of ice-cold 1,1,1-trichloroethane, and sucked dry as quickly as possible. 40

The crystals must be stored in an air-tight container, preferably sealed in a glass ampoule, as they react with moisture in the air.

1 Draw a labelled diagram of the refluxing reaction mixture (line 19). [5]

2 Here are the results of one experiment:

mass of iodine = 4.86 g
initial mass of tin = 4.27 g
final mass of tin = 3.14 g
relative atomic masses: Sn = 118, I = 127

Use these results to find the formula of the tin iodide, clearly showing the steps in your working. What is the percentage of tin in the compound? [5]

3 What safety precautions would you take whilst carrying out this experiment? [2]

4 Why will a small amount of product adhering to the tin not affect the calculated formula of the tin iodide? [2]

5 What type of chemical bonding is found in the product? Justify your answer from evidence in the passage. [3]

6 Why do you think that, during the reaction, the condensed liquid is purple, but when reaction is complete it is orange? [2]

7 Write a balanced equation for the reaction which occurs between tin iodide and water (line 43). [2]

8 Why should *small* amounts of *ice-cold* solvent be used to wash the product (lines 40–41)? [3]

Total = 25

Exercise 2·3

Liquid water—the story unfolds

hydrogen bonding, mechanism, solvation

It is incredible that research into the structure of such a simple substance as liquid water should still be a happy hunting ground for chemists, physicists and biologists! The monomer H_2O is now 'fully' understood and the wide range of different ice structures presents few remaining puzzles. However, liquid water is a different story: our understanding is still at an elementary level, and constantly under review. 5

Water molecules are effectively 'tetrahedral' since each H_2O molecule has two O—H bonds and two lone pairs (LPs) which can be used for hydrogen bonding. In ice each molecule is linked to four neighbours via two donor and two acceptor H-bonds. For liquid water, however, there are three inter- 10 linked questions:

• how many H-bonds are formed per molecule?
• how strong and how linear are such bonds? 15
• what are the lifetimes of these bonds?

Theories range from one extreme—that effectively all mole- cules are successful at forming four H-bonds all the time, however strong or weak—to the other extreme, that at any given moment some molecules are entirely successful, while 20 others fail completely.

Whatever model is suggested it is important that it makes chemical sense. It would seem more probable, therefore, that the truth lies somewhere between these extremes. A good model must allow H-bonds to range from being linear and at 25 least as strong as those found in ice at 0 °C, to being long, weak and non-linear. However, given that many bonds are very weak, it seems unavoidable that some will break:

$$H_2O_{bulk} \rightleftharpoons OH_{free} + LP_{free}$$

'Bulk' refers to four-H-bonded molecules (Figure 1, (a)), 30 OH_{free} and LP_{free} to three-H-bonded molecules ((b) and (c) respectively); in pure water the concentrations of the latter two species must be equal.

A key issue is the number of H-bonds broken at a given temperature. This is surely far greater than the number of 35 autoionisation events, and so must have some importance; however, even this is debated.

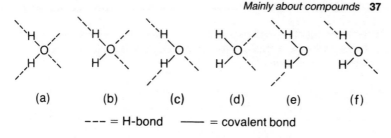

(a) (b) (c) (d) (e) (f)

--- = H-bond   —— = covalent bond

Figure 1 The possible arrangements of H-bonds around a water molecule.

We should now consider molecules with only two H-bonds. There are three possibilities (Figure 1, (d) (e) and (f)) but the third one is far more probable than the other two. This is 40
because in (d) and (e), the two H-bonds are tending to oppose (weaken) each other, whereas in (f) they reinforce (strengthen) each other.

Within the normal temperature range, one would expect structures (b), (c) and (f) to be significant entities, and that 45
four-H-bonded water would be inert. Reaction requiring attack by OH will depend on OH_{free} (Figure 1 (b) and (f)), whereas those requiring nucleophilic attack by a lone pair depend upon LP_{free} ((c) and (f)). This concept is simply an extension of the classes of reactions requiring H_3O^+ or OH^- respectively—the 50
neutral units are less reactive than the ions, but are present in far higher concentrations. Similarly, the solvation of ions depends upon these 'free' units. Cations are thought to solvate via LP_{free} molecules, whereas anions form strong, linear H-bonds with OH_{free} groups. 55

Adapted from an article with the same title by M. C. R. Symons,
Chemistry in Britain, *May 1989*

1 Explain what is meant by 'the monomer' of water (line 3). [2]

2 What is donated or accepted when a hydrogen bond forms (lines 11–12)? [2]

3 What does 'linear' mean in this context (line 15)? [3]

4 What two conditions are necessary for hydrogen bonding to occur between two molecules? [4]

5 Write an equation to show the autoionisation of water (line 36). Explain why the author believes this to be far *less* likely than the reaction given at line 29. [5]

6 Explain, in your own words, the meaning of the word 'solvation' (line 52). Illustrate your answer, with diagrams, by reference to Na^+ and Cl^- ions. [7]

7 Explain why the two H-bonds in structures (d) and (e) (Figure 1) oppose each other, but those in (f) strengthen each other. [7]

Total = 30

Additional work

Water is considered to be an atypical liquid, and many of its unusual properties are attributed to its highly H-bonded structure. Give an account of these properties.

Binary oxides

*acid–base reactions, lattice enthalpy, Periodic Table,
solubility, spectra*

With the exception of the first four noble gases, oxygen forms
binary compounds with every element. Because of its high
ionisation energy and strong attraction for electrons, oxygen
almost invariably gains electrons in its reactions: and thereby
oxidises the element with which it reacts. Metals, with their   5
low ionisation energies, generally form ionic oxides in their
lower oxidation states. Such oxides often have high lattice
energies, since the metal cations pack efficiently into the
interstices (vacant spaces) between the larger oxide ions, and
the melting points are typically high. Many metal oxides find   10
uses as high temperature materials, such as firebrick and
ceramics, whilst other uses depend upon their thermal stability as in, for example, the use of Y_2O_3 in fluorescent lights and
television tubes. This oxide emits a reddish glow when excited.

In general, the solubilities of metal oxides are low, but they   15
are significantly soluble when the lattice energy is small, as
happens when the cation is large and its charge is low. When
they dissolve, the oxide ions reacts with water to form hydroxide ions:

$$O^{2-}_{(aq)} + H_2O_{(l)} \rightarrow 2OH^-_{(aq)}$$ 20

In acids, water and the metal salt are formed.

Non-metals have sufficiently high electronegativities to prevent oxygen from removing electrons to form O^{2-} ions. Instead,
these elements form molecules containing polar covalent bonds
with oxygen to give gases, liquids, or low melting-point solids.   25
These oxides are acidic, particularly in higher oxidation states.
For example, chlorine(VII) oxide readily dissolves in water to
form perchloric acid:

$$Cl_2O_{7(l)} + H_2O_{(l)} \rightarrow 2HClO_{4(aq)}$$

Metal oxides where the metal has a high oxidation state are   30
also acidic and are best described as covalent (see the table).
Two simple rules are useful in deciding the acid–base character of an oxide:

1. For a given element, the acidity of the oxide increases with increasing oxidation state. For example, MnO dissolves   35
readily in acid, whereas Mn_2O_7 is water soluble to give an acidic solution.

2. For a given oxidation state, the acidity of an oxide increases with increasing electronegativity of the element. For example, SeO_2 is less acidic than SO_2.   40

Oxides of the main group elements in their highest oxidation states. Acid character increases Group I to Group VII, and basic character increases Period 2 to Period 6. The heavy line divides metallic and non-metallic elements

Period \ Group	1	2	3	4	5	6	7
2	Li_2O	BeO	B_2O_3	CO_2	N_2O_5	–	F_2O
3	Na_2O	MgO	Al_2O_3	SiO_2	P_2O_5	SO_3	Cl_2O_7
4	K_2O	CaO	Ga_2O_3	GeO_2	As_2O_5	SeO_3	Br_2O_7
5	Rb_2O	SrO	In_2O_3	SnO_2	Sb_2O_5	TeO_3	I_2O_7
6	Cs_2O	BaO	Tl_2O_3	PbO_2	Bi_2O_5	PoO_3	At_2O_7

These rules are best seen in operation in the table, in which the oxides of the main group elements in their highest oxidation states are shown with the acid–base trends.

Certain oxides, generally those of elements which lie on or near the diagonal metal/non-metal borderline of the Periodic   45
Table (see the table), are insoluble in water but will dissolve in both acidic and alkaline solutions. Such oxides are said to be **amphoteric.** For example, aluminium oxide (Al_2O_3) will dissolve in both HCl and NaOH solutions, in the latter case forming the aluminate anion, which may be written AlO_2^-.   50

Adapted from Chemistry, the Central Science, *T. L. Brown and H. E. LeMay,*
Prentice Hall (1988)

1 Give the meaning of the following terms:
   a binary (line 2),
   b lattice energy (lines 7–8),
   c thermal stability (lines 12–13), and
   d electronegativities (line 22).   [8]

2 Why do metals have low ionisation energies (line 6), and why should this result in the formation of ionic oxides? [4]

3 Explain in electronic terms why Y_2O_3 emits a glow when excited (lines 13–14). [3]

4 'Almost invariably' (line 4) implies some exception(s). Can you give one, justifying your choice? A careful examination of the table will help you to answer this. [2]

5 By a consideration of the energy changes involved, explain the relationship between **lattice energy** and **solubility** outlined in the second paragraph. [5]

6 Write balanced equations to illustrate
    a the reactions of the two manganese oxides (lines 35–7) with acid and with water, respectively, and
    b the acid and base behaviour of aluminium oxide (lines 48–50). [8]

Total = 30

The weathering of rocks

acid–base reactions, geochemistry, hydration, solubility

The solubility of different elements in water is clearly an important factor in the chemistry of weathering. For many elements, solubility is a function of the pH and the presence of anions such as carbonate, but a useful general guide to aqueous solubility of an element at neutral pH is provided by its **ionic potential**, defined as the formal charge, Z, divided by the ionic radius, r. A plot of charge against ionic radius for several elements with common positive oxidation states appears thus:

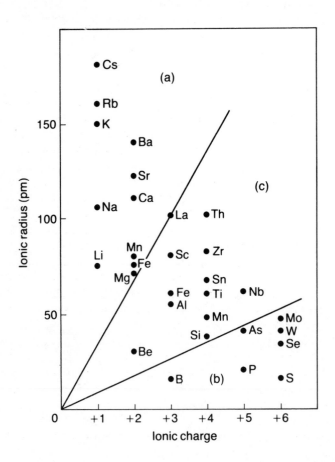

The heavy lines through the origin separate three groups. 10
Ions with $Z/r < 0.03$ pm^{-1} (a) are generally soluble at a pH
close to 7, forming simple hydrated cations. Those with
$Z/r > 0.12$ pm^{-1} (b) form soluble oxyanions; some of the elements in this category are non-metals, for which the concepts
of 'ionic charge' and 'ionic radius' have only a formal signi- 15
ficance. The least soluble elements at neutral pH are those
with intermediate Z/r values, occurring between the two lines
(c).

These elements form oxy and hydroxy species, which are
often polymerised and highly insoluble, although they may 20
dissolve more readily under acidic or basic conditions; for
example, silicon is more soluble at high pH, and amphoteric
elements, such as aluminium, can dissolve in both acidic and
alkaline solutions. It can be seen that the position of some
elements also depends on their oxidation states. Iron and 25
manganese are relatively soluble in the +2 state, but the
oxidised forms Fe^{3+} and Mn^{4+} are highly insoluble.

Using these general guidelines, one can understand the
kinds of chemical changes which take place in rocks during
weathering. The changes are shown most clearly under ex- 30
treme conditions, for example, in tropical climates. Alkali
metals, and those of Group II (except Be) will be washed out of
the rocks, as will many of the non-metals which form anions
(this includes, of course, the halogens, particularly chlorine
and bromine which occur as very soluble simple anions). Iron 35
and manganese present as 2+ ions (which is the case in many
igneous rocks) will also go into solution, although subsequent
oxidation by air will result in their precipitation. Some elements, on the other hand, will remain as solids. Minerals
which are unaltered by weathering may include SiO_2 (quartz), 40
oxides of Ti and Sn, and the lanthanide mineral monazite,
$LnPO_3$. These are classed as **resistates**. The ultimate dissolution of clay minerals produces **hydrosylates**, which include
the very insoluble element aluminium as the hydrated oxide
such as $AlO(OH)$. 45

It can thus be seen that there may be considerable sim-
plification of the rock chemistry, with the loss of soluble
elements leaving others separated as simple compounds. Many
important mineral deposits have been formed in this way,
including **bauxite**, made up of hydrated forms of aluminium 50
oxide, and iron ores such as **haematite** (Fe_2O_3), which can
result from the oxidation of Fe^{2+} released in weathering. Other

elements, such as titanium, zirconium, and the lanthanides, are often found in conjunction with silica.

The fact of soluble elements removed from rocks during weathering depends on their relative solubility, especially in conjunction with the anions also present, such as carbonate and sulphate. Many of these elements pass into the ocean. Some of these, such as calcium, precipitate fairly easily, especially as carbonates which form such widespread sedimentary rocks. Others, such as sodium, remain in solution for a very long time, although they eventually form **evaporite** deposits where enclosed seas or lakes are dried up by evaporation.

Adapted from The Elements, *P. A. Cox, Oxford (1989)*

1 Why is water solubility an important factor in weathering? [2]

2 Use the graph to calculate the ionic potential of caesium. What is another name for 'ionic potential'? [4]

3 Explain why silicon (line 22) is more soluble at high pH (it will normally occur as its oxide). [4]

4 Using examples from the text or the graph, and giving equations, formulae, and/or diagrams as appropriate, explain the meaning of
   a simple hydrated cations (line 12),
   b oxyanions (line 13), and
   c amphoteric elements (lines 22–3). [8]

5 Why should the more oxidised forms of these elements be less water soluble (lines 25–7)? [3]

6 Bauxite (line 50) normally contains silica (SiO_2) and iron(III) oxide as impurities. Explain briefly why this should be so. [4]

7 What oxyanions (one each) might molybdenum (Mo) and arsenic (As) form? [2]

8 Name *three* other elements which are likely to be found in evaporite deposits (lines 62–3). [3]

Total = 30

Exercise 2·6
Molecules in space

absorption spectra, energy transformations, formulae, polarity

At the very high temperatures needed to synthesise elements in stars, no molecules can exist: indeed, most matter is present as free nuclei and electrons. In the cooler (around 3000 K) outer layers of giant stars, however, some diatomic molecules have been detected by observation of the absorption spectra of   5
these stars. In interstellar space, the density of matter is around 10^6 atoms per m^3, mostly hydrogen and helium. Although the temperature is low enough for the formation of chemical bonds, any molecules which *did* form would rapidly be dissociated by high-energy photons (ultraviolet radiation)   10
from stars, and so it was once believed that none but the simplest molecules could exist in space. However, there are clouds of relatively high density—up to 10^{10} atoms per m^3— which absorb UV quite strongly. Any molecules which form deep within these clouds are thus protected, and to date   15
around 50 molecular species with up to 13 atoms have been detected in the regions between stars.

These molecules must have been formed by collisions be-tween atoms and/or other molecules. The fastest reactions are those involving ions, since ions will polarise other species,   20
inducing a dipole moment and leading to a long-range attrac-tion. For example:

$$C^+ + H_2 \rightarrow CH^+ + H$$

Molecular ions like CH^+ can undergo further reactions; this is thought to be the major route to organic molecules in interstel-   25
lar space.

If a chemical bond is to form as the result of a collision, then the energy of the colliding species must be dissipated rapidly, otherwise they will just bounce off one another. In the case of molecular collisions, one way this can be accomplished is by   30
rapidly transferring the energy to various bond vibrations and rotations ('internal' energy), the energy eventually being lost by infrared and microwave photon emission. The more complex the species, the more easily this process can occur. We can picture the collision between A and B thus: 35

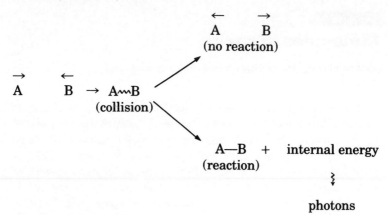

This cannot, however, explain the formation of a bond between two hydrogen atoms, by far the most common reaction in interstellar space. Hydrogen atoms are unable to convert their translational energy into internal energy in this way, and hence another mechanism is needed to explain how their collision energy is dissipated. Three possibilities have been suggested: 40

1. A simultaneous collision with a third species, M, which then carries away the energy: 45

$$\vec{H} \quad \overset{\leftarrow}{H} \rightarrow H\!\sim\!\!H \rightarrow H\!-\!H$$
$$\underset{M}{\uparrow} \qquad M \qquad \underset{\downarrow}{M}$$

even in relatively 'dense' clouds, however, such collisions would be rare and cannot explain the frequency of dihydrogen formation.

2. The direct emission of a photon immediately after collision: 50

$$\vec{H} \quad \overset{\leftarrow}{H} \rightarrow H\!-\!H + \text{photon}$$

Since the collision will last only 10^{-13} s, however, this is again thought an unlikely explanation, since there is not enough time for the photon emission and bond formation before the atoms move apart. 55

3. The most likely mechanism is that the two atoms are adsorbed onto the surface of interstellar dust particles, lose their energy, combine, and then are desorbed as the molecule:

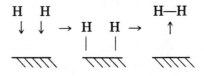

60

Adapted from The Elements, *P. A. Cox, Oxford (1989)*

1 Express the atomic concentration in dense interstellar clouds (line 13) in mol dm^{-3} (Avogadro's constant $= 6{\cdot}00 \times 10^{23}$). [3]

2 Can you suggest why there are no molecules and very few atoms to be found in the interiors of stars (lines 1–3)? [3]

3 What are absorption spectra, and how can they be used to detect molecules in this way (lines 4–5)? [4]

4 By what mechanism might interstellar clouds absorb UV photons (lines 13–14)? [3]

5 Explain the meaning of
a polarise (line 20);
b dipole moment (line 21).
Illustrate your answers with examples from the text. [6]

6 Why can hydrogen atoms not absorb collision energy (lines 39–40)? [2]

7 In the equation given on line 46, assume M to be a water molecule. Give, with suitable diagrams, some of the ways that it might absorb the energy of the collision. [4]

8 As well as those mentioned in the text, the atoms found in interstellar space include oxygen, nitrogen, and sulphur. Give the names and formulae of *five* uncharged simple molecules which might be detected there. [5]

Total = 30

Exercise 2·7

The Earth's atmosphere and the development of life

biochemistry, environment, radioactivity, noble gas, photochemistry

Apparently, the heavier planets still retain their original atmospheres, but the atmosphere of the Earth, because of the planet's low gravitational field and consequent inability to retain light gases, seems to have its origin in gases released from within the planet. For example, the abundances of the noble gases in the Earth's atmosphere are markedly different from their solar abundances, since their inability to combine with other elements meant that they were soon lost to space. All the helium and argon present today is due to radioactive decay processes; argon from potassium 40, and helium from uranium and thorium. These two gases are being produced at a roughly equal rate, as shown by their abundances in natural gas, but argon is very much more abundant in the atmosphere. 5

It is believed that the early atmosphere of the Earth consisted mainly of nitrogen, carbon dioxide, and water vapour. Subsequent evolution of this depended upon many factors. Minor constituents such as methane and ammonia would have been rapidly destroyed by solar radiation-induced photochemical reactions. Almost all the carbon dioxide would have been deposited as sedimentary rocks via solution in the sea and the action of creatures such as molluscs. Today, the sea contains 50 times more carbon dioxide than the air, and 10^5 times more is locked up in rocks. Small amounts of oxygen would have been produced in the early atmosphere by the photochemical decomposition of water, followed by the escape of atomic and molecular hydrogen into space. This, however, does not explain the large amount of oxygen in the present day atmosphere: if this photochemical route were the only source of oxygen, then the concentration would be around 10^{-12} of its actual value. To account for this, and the presence of traces of methane and ammonia, we must turn to biological processes— the development of life. 15 20 25 30

Synthesis of fairly complex organic molecules is a necessary precursor of life. In a famous series of experiments, mixtures of gases similar to the early atmosphere were subject to spark discharges and ultraviolet radiation (as the early atmosphere 35

would have been), and small amounts of amino acids and other relatively simple pre-life compounds were formed. Subsequent experiments have shown how these could have been condensed into polypeptides, mono- and poly-nucleotides, probably via   40
absorption onto clays. Study of ancient rocks reveals that life, in the form of microbial organisms, started at least 3.8×10^9 years ago (the Earth itself is 4.6×10^9 years old). Later, blue-green algae appeared; it seems probable that these organisms were the first to develop photosynthesis. They can exist in   45
either anaerobic or aerobic conditions, and can split either hydrogen sulphide or water to release sulphur or oxygen respectively. The latter process caused the oxygen concentration in the atmosphere to rise, which meant that the organisms had to develop a tolerance to this very reactive gas.   50

Another important consequence of this increased oxygen level was that life could for the first time move onto dry land. Living organisms are damaged by ultraviolet radiation, so early life forms could only survive in a depth of at least 10 m of water, since this provided protection from the sun. However,   55
dioxygen and its photochemical allotrope ozone between them absorb ultraviolet radiation, so a whole range of land plants could now evolve, and the atmospheric oxygen concentration continued to rise until it was more or less what it is today.

When living organisms die, however, decay processes even-   60
tually convert all the carbon to carbon dioxide, and there is no net gain in dioxygen. Were it not for the fact that much of this dead biomass does not decay—it becomes buried and meta- morphosed—there would be much less dioxygen in the air. If all the carbon in this buried matter is oxidised to carbon   65
dioxide, then the oxygen level in the atmosphere could become much closer to its early, pre-life, value.

Adapted from Origin and Evolution of the Atmosphere, *R. P. Wayne,*
Chemistry in Britain, *March 1988*

1 Write nuclear equations to show how helium and argon are formed (lines 9–11). Use a Periodic Table for any figures that you need.
Why should argon be the more abundant of these two gases in the Earth's atmosphere (lines 11–13)?   [6]

2 Giving relevant equations, show how carbon dioxide in the air might be converted into a *named* sedimentary rock (lines 19–21).   [5]

3 What biological processes might account for the presence of methane and ammonia in the atmosphere (lines 30–32)?   [2]

4 Write a displayed formula of *one* amino acid (line 37), and an equation to show how *two* of these acid molecules condense on the way to becoming a polypeptide (line 40).   [4]

5 Explain what is meant by the terms
  a photochemical reaction (lines 18–19);
  b photosynthesis (line 45).
  Use examples from the text to write illustrative equations (one each).   [8]

6 Explain how the conditions used in the experiments outlined in lines 34–6 were similar to those found in the early atmosphere.   [5]

Total = 30

Additional work

What is meant by 'buried and metamorphosed' in this context (lines 63–4)? How are humans completing this process of 'decay', and what consequences might this eventually have?

Exercise 2·8

An alternative to the Haber process?

biochemistry, coordination chemistry, industry, Lewis structures, world food problem

Nitrogen fixation is the process that converts atmospheric nitrogen (dinitrogen) into ammonia or oxides of nitrogen. It is essential for life on this planet; without this continuous supply of 'fixed' nitrogen to the soil, reserves would rapidly be used up by natural return of dinitrogen to the air, by microbial denit-   5
rification of nitrate. The amounts of nitrogen fixed each year are about 5×10^5 t by the chemical industry, and almost twice this by natural means (microbes and thunderstorms).

The major industrial source of fixed nitrogen is the Haber process, which has been operating successfully for about 80   10
years. It is highly developed and capital intensive, with a large energy and raw material consumption. Moreover, because of the centralised nature of the factories, transport costs to the user are high. Biological conversion, on the other hand, occurs at ambient temperature, and uses the sun directly as an   15
energy source. If a practical way of imitating this could be found, it could be done locally in cheap-to-run 'low tech' factories. With the world population expected to reach 8×10^9 early next century, the importance of understanding the bio-logical conversion of nitrogen is clear, and scientists have long   20
been considering this problem.

Biological fixation is performed by two types of bacteria: symbiotic, which live in association with plants (e.g., in root nodules), and free-living in the soil. Dinitrogen gas is very unreactive, it cannot be used directly by plants. The bacteria   25
have to find a way of making it accessible: they do this by reducing it to ammonia, catalysing the process with nit-rogenase enzymes which are highly complex protein structures containing sulphur, iron, and either molybdenum or vana-dium. These enzymes can be extracted from the bacterial cells   30
and can reduce dinitrogen to ammonia in the laboratory in the presence of an electron source such as magnesium adenosine triphosphate (Mg-ATP).

The vital role of molybdenum in nitrogen fixation has long been known, but vanadium has only more recently been   35
recognised as an alternative. Essentially, the Mo or V atom is square-planar tetracoordinated at the centre of a protein

molecule (either by one tetradentate or two bidentate ligands, the donor atoms being sulphurs). The Mo can then coordinate with two dinitrogen molecules to form an octahedral hexacoordinated dipositive ion: 40

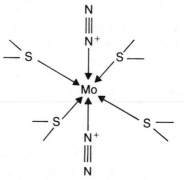

The reduction then follows a series of steps involving this complex ion, protons, and the electron source (such as Mg-ATP), and may be summarised stoichiometrically as:

$$N_2 + 10H^+ + 8e^- \rightarrow 2NH_4^+ + H_2$$ 45

Carbon monoxide, isoelectronic with dinitrogen, is not reduced by nitrogenase, but it does inhibit the action of the enzyme, presumably by binding irreversibly to it.

Synthetic analogues of the enzyme have been prepared, and the dinitrogen complexes made from them (at ambient 50
temperature and pressure) have been reduced electrochemically in acidic solution to yield ammonia. It is conceivable that a system based on this chemistry, using the sun, wind or tide as an energy source, could be developed as an alternative to the Haber process. 55

Adapted from an article by R. L. Richards in Chemistry in Britain,
February 1988

1 What role do thunderstorms (line 8) play in the fixation of nitrogen? [3]

2 Give an equation and the operating conditions for the Haber process itself (lines 9–10), and describe briefly, with relevant equations, how the reactants are obtained from the raw materials. Show how these justify the author's statement that the process has a large energy and raw material consumption. [10]

3 Draw a Lewis structure (dot-and-cross diagram) of the dinitrogen molecule (line 24), and use this to explain
  a why it is very unreactive, and
  b how it bonds to a molybdenum atom (lines 39–41).   [7]

4 Using examples taken from the passage, explain the terms
  a isoelectronic (line 46);
  b tetradentate (line 38).   [4]

5 Explain what is meant by 'reduced electrochemically' in this context (lines 51–2).   [2]

6 Why is a source of energy needed for the enzyme process? Describe briefly how any *one* of the sources mentioned might be used (lines 53–5).   [5]

7 What does the author believe to be the advantages of this method of nitrogen fixation, particularly when compared to the Haber process?   [4]

Total = 35

Additional work
Why could it become important, in terms of the world's finite resources and growing population, to find an alternative to the Haber process?

Practical applications of coordination complexes

colorimetric analysis, displayed formulae, environment, K_{sp}, Le Chatelier's principle, Lewis structures, photography, safety

When a transition metal forms a stable coordination complex, the metal itself becomes 'hidden' by the surrounding ligands, and profound changes in the physical and chemical properties of the metal often result. This fact has given rise to many important applications, a selection of which is outlined below. 5

1. Gold is normally resistant to atmospheric oxidation—witness the many untarnished gold artifacts of great antiquity—but in the presence of the cyanide ion, it is easily oxidised due to the formation of a stable, water soluble cyano complex: 10

$$4Au_{(s)} + 8CN^-_{(aq)} + O_{2(g)} + 2H_2O_{(l)} \rightarrow 4Au(CN)^-_{2(aq)} + 4OH^-_{(aq)}$$

This reaction forms the basis of the extraction of native gold from low-grade ores.

2. Complexing agents also enhance the solubility of metal salts. For example, AgBr is decomposed to its elements by 15 exposure to light and is thus crucial to photography. The compound is insoluble in water, but dissolves readily in the presence of the thiosulphate ion:

$$AgBr_{(s)} \rightleftharpoons Ag^+_{(aq)} + Br^-_{(aq)} \quad K_{sp} = 7 \cdot 7 \times 10^{-13}$$

$$Ag^+_{(aq)} + 2S_2O_3^{2-}_{(aq)} \rightleftharpoons [Ag(S_2O_3)_2]^{3-}_{(aq)} \quad K = 1 \cdot 6 \times 10^{13}$$ 20

Sodium thiosulphate decahydrate, known as hypo, is used in black-and-white photography to dissolve unchanged AgBr from exposed film, leaving the photographic negative.

3. Complex formation can also be used in analysis. For example, Fe^{3+} cations in aqueous solution react with thiocyanate 25 ions:

$$[Fe(H_2O)_6]^{3+}_{(aq)} + SCN^-_{(aq)} \rightarrow [Fe(H_2O)_5SCN]^{2+}_{(aq)} + H_2O_{(l)}$$

The hydrated cation is a brown colour, whereas the thiocyanate complex is a deep blood-red: this can be used as a qualitative test for Fe^{3+}, or colorimetrically to measure the 30 concentration of iron in solution.

4. Carbon monoxide is able to form very strong coordinate bonds, due to the unique electronic structure of its carbon atom: this accounts for its extremely poisonous nature, as it binds to the haeme molecule in blood much more ⁣35 strongly than oxygen can. Indeed, carbon monoxide will even coordinate with metals in the zero oxidation state. This discovery led to the development of the Mond process for the purification of nickel. A stream of CO is passed over the hot metal: ⁣40

$$Ni + 4CO \rightarrow Ni(CO)_4$$

The very volatile nickel carbonyl separates from the impurities, and the reaction is reversed elsewhere by stronger heating to give the pure metal. The carbon monoxide is recycled.

5. Although transition metals form the most stable complexes, ⁣45 other metals will also coordinate more weakly with ligands. Ca^{2+} and Mg^{2+} ions do not precipitate with detergent molecules as they do with soap, but they do complex with them, and hence render the detergent less effective in hard water. Polyphospate ions will complex with these metals more ⁣50 strongly, and so sodium polyphosphate, $Na_5P_3O_{10}$, is added to washing powders to render the Group II cations inactive. Polyphosphates, however, are thought to be environmentally harmful—they can lead to the excessive growth of algae in rivers, for instance—and so the use of polyphosphates is being ⁣55 phased out; 'green' washing powders are phosphate-free.

In part adapted from Chemistry, The Central Science, *T. L. Brown and*
H. E. LeMay, Prentice Hall (1988)

1 What hazard would be associated with the extraction of gold using the process outlined in 1? [1]

2 What does the word 'native' mean in this context (line 12), and what is the oxidation number of gold in the complex given in line 11? [2]

3 What remains on the photographic negative after the AgBr has been dissolved? Explain the mechanism by which it was formed, giving a relevant equation (lines 21–3). [3]

4 Draw Lewis structures (dot-and-cross diagrams) of the following ions:

   a cyanide (line 8),

   b thiosulphate (line 18), and

   c thiocyanate (line 25).

   Indicate on your diagrams which electrons are involved in the coordinate bond formation. [9]

5 How are these cations (line 52) rendered inactive? Draw a displayed formula of a polyphosphate ion (line 50), indicating clearly the formal location of the charges. [4]

6 What name is given to K_{sp} (line 19)? Write an expression to define it. Examine the two equilibria (lines 19 and 20), and by means of Le Chatelier's principle explain precisely how the addition of thiosulphate increases the solubility of AgBr. [7]

7 Why should excessive growth of algae be undesirable (line 54), and what does the word 'green' mean in this context (line 56)? [4]

Total = 30

Additional work

Give a brief account of the principles of colorimetric analysis (line 30).

Compounds with boron–nitrogen bonds

bonding, electronic structure, equations, formulae,
intermolecular forces, mechanism

Because boron has one less electron than carbon, and nitrogen
has one more, many B—N compounds are known which are
isoelectronic with their C—C analogues. Although the former
have similar shapes and bonding, they are very different to
the carbon compounds in their chemistry, mainly due to the   5
electronegativity difference between the two elements. The
simplest binary compound, boron nitride, is prepared by heating boron oxide with ammonia at 1200 °C:

$$B_2O_{3(l)} + 2NH_{3(g)} \rightarrow 2BN_{(s)} + 3H_2O_{(g)}$$

Boron nitride has a planar structure of edge-shared hexagons   10
of alternating boron and nitrogen atoms. The distance between
the sheets is 0·333 nm (compare this with the equivalent
distance of 0·335 nm in graphite), and the bonding is weak;
hence boron nitride, like graphite, can be used as a lubricant.
Here the similarity with graphite ends. In the stacking of the 15
layers, B and N atoms alternate vertically (see diagram below),
and hence the rings are directly over each other: this is a
consequence of the partial positive charge on the boron and the
partial negative charge on the nitrogen. In graphite the hexagons are staggered. Another striking difference is that boron   20
nitride is a colourless electrical insulator.

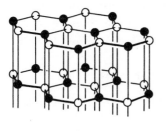

● B   ○ N

Boron nitride   Graphite

Compounds isoelectronic with saturated hydrocarbons, known as aminoboranes, may be prepared by the reaction of amines or ammonia with diborane, the simplest being the analogue of ethane: 25

$$B_2H_{6(g)} + 2NH_{3(g)} \rightarrow 2H_3B{-}NH_{3(s)}$$

Physically the two are very different; non-polar ethane boils at $-89\,°C$, whereas the boron nitride compound has a high dipole moment, and is a solid with a low vapour pressure at room temperature.   30

Several analogues of carboxylic acids are known, such as ammonia carboxyborane, H_3NBH_2COOH. Some of these compounds display significant physiological activity, including tumour inhibition and reduction of serum cholesterol in certain cases.   35

Many unsaturated boron nitride compounds are known. The simplest is isoelectronic with ethene, but only has a transient existence in the gas phase, since it readily forms a cyclic trimer. However, if the double bond is protected from attack by two bulky alkyl groups, and by two chlorines on the boron, a   40
stable monomer is formed. Such compounds can be prepared by the reaction between a dialkylamine and boron trichloride.

Borazine, $B_3N_3H_6$, is isoelectronic and isostructural with benzene, but the two compounds have little chemical resemblance. The former does not have the same delocalised π 45
structure as benzene, since the π electrons are concentrated on the nitrogen atoms, and hence it can be regarded as having three polar, reactive, double bonds. A typical reaction is the electrophilic addition of three molecules of HCl to form a trichlorocyclohexane analogue.   50

Many symmetrically trisubstituted derivatives of borazine have been prepared; for example, ammonium chloride reacts with boron trichloride in hot chlorobenzene solution to form *B*-trichloro-borazine:

$$3NH_4Cl + 3BCl_3 \rightarrow Cl_3B_3N_3H_3 + 9HCl \qquad 55$$

Alkyl ammonium chlorides yield the equivalent *N*-trialkyl compounds.

Adapted from Inorganic Chemistry, *D. F. Shriver, P. W. Atkins and C. H. Langford, Oxford (1990)*

1 What *type* of inter-plane bonding is found in graphite and in boron nitride? Why is the electrical conductivity of the two substances so different? [4]

2 Give the electronic structures of boron ($Z = 5$) and nitrogen ($Z = 7$), and use them to explain precisely the type of bonding formed between these two elements. [7]

3 By discussing the intermolecular forces involved, explain the difference in boiling points between ethane and its boron nitride analogue (lines 27–30). [3]

4 What is the name of the carbon analogue of ammonia carboxyborane (line 32)? [2]

5 Write an equation, using displayed formulae, for the polymerisation mentioned in lines 38–9. [3]

6 Using R to represent an alkyl group, write an equation for the reaction between a dialkylamine and boron trichloride (lines 41–2). [3]

7 Write an equation for the addition of HCl to borazine (lines 48–50). [4]

8 What do *B*- (line 54) and *N*- (line 56) mean? Draw a displayed formula of *N*-trimethyl-*B*-trichloro-borazine (line 55) to illustrate your answer. [4]

Total = 30

ORGANIC CHEMISTRY

The preparation and use of Grignard reagents

bond strength, environment, equations, mechanism, organometals, synthesis

The first **organometallic compound**—that is, a compound containing a metal–carbon bond—was made in 1849, when Frankland prepared the volatile, spontaneously flammable diethyl zinc. Since then, organometallic compounds of many metals have been prepared, ranging from solid, highly reactive   5
potassium alkyls to the only industrially important one, tetraethyl lead, used as a petrol additive but now being phased out. The methods of preparing such compounds usually involve replacing a halogen atom by a metal, for example, the reaction of lithium with iodopropane: 10

$$C_3H_8I + 2Li \rightarrow C_3H_8Li + LiI$$

In all organometallics, the metal–carbon bond is highly polar, or even ionic, and hence these compounds are useful reagents in organic syntheses, as they contain a strongly nucleophilic carbon atom.   15

Grignard reagents, RMgX (X = a halogen), were developed by Victor Grignard at the beginning of this century, for which work he won the Nobel prize in 1912. They are prepared by refluxing magnesium turnings with a halogenoalkane or -arene in ethoxyethane solution, in absolutely dry conditions: 20

$$RX + Mg \rightarrow RMgX$$

The preferred halogen is bromine. In the case of iodoalkanes, the Grignard reagent first formed reacts with a further molecule of the iodoalkane, to produce the hydrocarbon R—R. Chloroalkanes react less readily, and chloroarenes not at all.   25
Grignard reagents can be used to synthesise many classes of compound, but they cannot be isolated, and so must be used as soon as they have been made. Essentially, a dry solution of the

other reagent in ethoxyethane is added to the Grignard solution, when an addition reaction occurs. The adduct is then 30
hydrolysed by adding water to give the desired product.

Primary alcohols are made by reaction of the Grignard reagent with methanal:

$$RMgBr + HCHO \rightarrow RCH_2OMgBr \xrightarrow{water} RCH_2OH + Mg(OH)Br$$

If epoxyethane is used in place of methanal, two CH_2 units are 35
introduced into the chain, the overall reaction being:

$$RMgBr + H_2C\!\!-\!\!CH_2 + H_2O \rightarrow RCH_2CH_2OH + Mg(OH)Br$$
$$\diagdown\;\diagup$$
$$O$$

Secondary alcohols are made by using any other aldehyde, R'CHO, and **tertiary alcohols** by using ketones, R'COR''.

Carboxylic acids may be synthesised by passing dry car- 40
bon dioxide into the solution of the Grignard reagent, followed by hydrolysis:

$$RMgBr + CO_2 + H_2O \rightarrow RCOOH + Mg(OH)Br$$

Alkanes are produced by simply adding water, or better a dilute acid, to the solution of the Grignard reagent. This is not 45
a particularly useful type of reaction but it could be employed to make a pure sample of a particular alkane.

1 What shape is the tetraethyl lead molecule? Explain why, on environmental grounds, this compound is being phased out as a petrol additive (lines 6–7). [2]

2 What is the direction of polarity of a metal–carbon bond, and when is it likely to be ionic (lines 12–13)? Explain your reasoning, giving an example of each type from the text. [5]

3 What reaction would occur if absolutely dry conditions were not used during the preparation of Grignard reagents (line 20)? [2]

4 Write balanced equations, including the hydrolysis step, for the preparation of secondary and tertiary alcohols by this method, using the general formulae given (lines 38–9). [4]

5 Write a balanced equation for the reaction between an iodoalkane, RI, and the corresponding Grignard reagent (lines 22–4). [2]

6 What is meant by 'nucleophilic' (line 14)? Illustrate your answer with a mechanistic equation, using the reaction of a Grignard reagent with water (lines 44–5). [4]

7 Why do you suppose that chloroalkanes react less readily with magnesium than do the bromo- and iodoalkanes, and chloroarenes do not react at all (line 25)? [4]

8 Draw structural formulae for diethyl zinc (line 4) and ethoxyethane (line 20). Can you offer an explanation, in terms of bond energies and kinetics, why the former is spontaneously flammable in air, but the latter is not? Sketch graphs to illustrate your answer. [7]

Total = 30

Additional work

Outline a synthesis of butan-2-ol, via a Grignard reagent, using ethanol as your *only* organic starting material. For each step in your synthesis, give the reagents and necessary conditions.

The preparation of 4-bromophenylamine

apparatus, experimental technique, mechanism,
% yield, recrystallisation, safety, synthesis

The aromatic ring of phenylamine is very electron rich, due to the ability of the lone pair on the nitrogen to be delocalised into the π-system, and consequently it is highly activated towards electrophilic substitution. Even with aqueous bromine in the cold, it undergoes polysubstitution to yield 2,4,6- 5 tribromophenylamine. However, 'tying up' the lone pair of the nitrogen by converting the phenylamine to an amide greatly reduces the activity of the ring so that it may be monosubstituted by bromine. The ring is still reactive enough, however, for the substitution to occur without a catalyst, unlike the 10 bromination of benzene itself. The bulky amide group also favours substitution at the 4-, rather than the 2-, position, leading to an increased yield of the desired product. After bromination, the amide group is hydrolysed back to the amine.

Preparation of *N*-phenylethanamide 15
Add 10 g of phenylamine to 25 cm^3 of ethanoic acid in a flask, followed by 12 g of ethanoic anhydride. Mix well, and allow the mixture to stand at room temperature for 5 minutes. Dilute with 100–200 cm^3 of water until crystallisation of the product occurs. When this is complete filter off the colourless, lustrous 20 crystals by suction, wash with cold water, dry in air, and record the yield.

Preparation of *N*-(4-bromophenyl)ethanamide
Dissolve 5·0 g of the *N*-phenylethanamide and 6·0 g of bromine in separate 25 cm^3 portions of ethanoic acid, then add the 25 bromine solution to the other one over 5 minutes, while stirring. Allow the mixture to stand at room temperature for 15 minutes and then pour into .300 cm^3 of cold water. Stir well, adding 1–2 g of sodium metabisulphite to remove any remaining bromine. Filter off the product by suction, wash with a 30 little cold water, dry, and record the yield.

Preparation of 4-bromophenylamine

Place 5·0 g of the *N*-(4-bromophenyl)ethanamide in a 100 cm^3 round-bottomed flask and add 50 cm^3 of 5·0 M hydrochloric acid. Fit a reflux condenser, boil the mixture until all the solid   35 dissolves, and then continue boiling for a further 10 minutes. Cool the solution in ice, and cautiously add sodium hydroxide until the solution is alkaline (use pH paper). The product separates as an oil which solidifies on cooling and scratching. When all has crystallised, filter by suction, wash with a little   40 cold water, and recrystallise from a mixture of water and ethanol. If the product is coloured, a little charcoal may be added to the hot solution during the recrystallisation, and then filtered off (keeping the solution hot). Finally, dry the crystalline product, record your yield, and determine the melting   45 point. The amine is susceptible to oxidation by the air, particularly when wet or in solution, but will keep indefinitely when dry.

Adapted from Experimental Organic Chemistry, *Harwood and Moody,*
Blackwell (1989)

1 Write a reaction scheme for the three stages of this preparation, noting any other products. [5]

2 Phenylamine is toxic by skin absorption, and a cancer suspect agent. Ethanoic acid and ethanoic anhydride are both corrosive and have irritating vapours. Bromine has a toxic, irritating vapour, and causes nasty burns if it comes into contact with the skin. Outline the safety precautions that you would take during this preparation. [4]

3 In the second stage of the preparation, 5·20 g of dry *N*-(4-bromophenyl)ethanamide were obtained. Calculate the percentage yield for this stage. What explanations can you offer for the fact that the yield was not 100%?
(C = 12, H = 1, O = 16, Br = 80, N = 14) [7]

4 Name the apparatus you would use when
  a adding the bromine to the phenylamine solution (lines 25–7), and
  b filtering by suction (line 30).
  Sketch the apparatus used in part **b**. [6]

5 What technique would you use to monitor the pH of the solution during the addition of the sodium hydroxide (lines 37–8)? [2]

6 Sodium metabisulphite (line 29) can be regarded as a solution of sulphur dioxide. Write a balanced equation showing how it removes the last traces of bromine. [3]

7 a Describe how you would recrystallise the product (lines 41–2).
  b What characteristics are important in the choice of a solvent for recrystallisation?
  c What is the purpose of determining the melting point of the product? [9]

8 Explain, with the aid of a suitable diagram, why the ring in the ethanamide derivative is less activated than that in phenylamine. [4]

Total = 40

Additional work

What catalyst is used in the reaction of bromine with benzene (lines 9–10)? Give a detailed mechanism for the reaction.

Exercise 3·3

The oxidation of alcohols

apparatus, experimental technique, safety, structure, yield

Oxidation of organic molecules is complementary to, and equally as important as, their reduction. Oxidation is essentially the reverse of reduction, and can involve the removal of two hydrogens, the addition of oxygen, or the replacement of hydrogen by a heteroatom functional group.   5

Although various organic molecules can be oxidised, the conversion of an alcohol to a carbonyl compound is perhaps the most frequently encountered oxidation process in organic synthesis, and many reagents have been developed for this important transformation. Primary alcohols are oxidised first   10
to aldehydes, but since aldehydes are themselves easily oxidised, the oxidation of primary alcohols often continues to form the carboxylic acid. However, by appropriate choice of reagent the oxidation can be controlled and stopped at the aldehyde stage. Secondary alcohols are readily oxidised to ketones, but   15
tertiary alcohols are not usually oxidised, although under acidic oxidising conditions they may dehydrate to alkenes, which may themselves be subject to oxidation.

One of the most commonly used oxidants is aqueous chromic acid, prepared from sodium dichromate and sulphuric acid. Its   20
use can be illustrated by the transformation of 2-methylcyclohexanol to 2-methylcyclohexanone. The oxidant is used in excess, and the reaction is carried out in a two-phase ether-water system at 0 °C. The work-up is simple, involving separation and washing of the ether layer, and the product   25
ketone is purified by distillation at atmospheric pressure.

Preparation of 2-methylcyclohexanone

Dissolve 10 g of sodium dichromate dihydrate in 30 cm^3 of water in a 100 cm^3 beaker and, whilst stirring rapidly, slowly add 7·4 cm^3 of concentrated sulphuric acid. Make up the total   30
volume of the solution to 50 cm^3 by adding water, place the oxidising solution in an icebath, and allow it to cool for about 30 minutes.

Meanwhile, place 5·7 g of 2-methylcyclohexanol and 30 cm^3 of ethoxyethane in a 250 cm^3 round-bottomed flask, add a   35
magnetic stirrer bar, and fit an addition (tap) funnel to the flask. Stir the solution, and place the flask in an icebath for 15

minutes. Keep the flask in the icebath, and add about half of the ice-cold oxidising solution *dropwise* to the rapidly stirred reaction mixture. Add the remaining oxidant slowly over about five minutes, and stir the mixture rapidly in the icebath for a further 20 minutes.

40

Stop the stirrer, and allow the layers to separate. Transfer the mixture to a separating funnel, and separate the layers. Extract the lower aqueous layer with $2 \times 15\,cm^3$ portions of ethoxyethane. Combine the three ether layers, and wash them with $20\,cm^3$ of aqueous sodium carbonate solution, and then with $4 \times 20\,cm^3$ portions of water. Dry the ether layer over anhydrous magnesium sulphate. Filter off the drying agent, and evaporate the solvent on the rotary evaporator. Transfer the residue to a $10\,cm^3$ flask and distil the liquid at atmospheric pressure, collecting the product at between 160 and 165 °C. Record the yield of your product.

45

50

Adapted from Experimental Organic Chemistry, *Harwood and Moody,*
Blackwell (1989)

1 Explain the following terms:
  a primary, secondary, and tertiary alcohols (lines 10–16);
  b heteroatom functional group (line 5);
  c two-phase (line 23). [6]

2 Write an equation for the oxidation reaction, using [O] to represent the oxidising mixture, clearly showing the structures of the organic reactant and product. [4]

3 What safety precautions would you take during the preparation of 2-methylcyclohexanone? [4]

4 Why is the product solution washed with aqueous sodium carbonate (lines 46–7)? What precaution would you take whilst carrying out this procedure, and why? [3]

5 Sketch and label the apparatus used
  a for the oxidation stage;
  b for the final distillation. [6]

6 Calculate the theoretical yield of 2-methylcyclohexanone. How would you check the purity of your product?
($C = 12, H = 1, O = 16$)
Name the reagent and conditions you would use to show that
  a it was a carbonyl compound, and
  b it was a ketone and not an aldehyde.   [7]

7 Describe the technique you would use, mentioning any precautions,
  a to extract the lower aqueous layer with ethoxyethane (lines 45–6);
  b to determine the yield of the final product (line 53).   [5]

Total = 35

Toxic combustion products

*displayed and empirical formulae, % composition,
polymers, radicals, safety*

Plastics and other synthetic polymers are a major source of
toxic and irritant gases during a fire. These gases include
carbon monoxide, hydrogen cyanide, hydrogen chloride, sulphur dioxide, nitrogen oxides, phosgene ($COCl_2$), isocyanates
(RNCO), and nitriles. Smoke particles may also be important   5
in the toxic effects, because small particles may be inhaled
deeply into the lungs, carrying absorbed irritants.

Carbon monoxide is formed by the incomplete combustion of
carbon-containing materials, and inhibits oxygen transport in
the blood by binding to haemoglobin more strongly than   10
oxygen can. Death occurs when over 50% of the blood haemoglobin is bound to CO, which requires levels of only a few
thousand parts per million (ppm). Hydrogen cyanide and
oxides of nitrogen are usually formed in fires of nitrogen-
containing polymers such as polyurethanes and polyacrylonit-   15
rile, which are widely used in domestic furnishings. Polyvinyl
chloride is the most common commercial thermoplastic and is
used in flooring, cables and pipes. Its thermal decomposition
produces about 50% by mass of hydrogen chloride at relatively
low temperatures (below 225 °C). Sulphur dioxide comes from   20
sulphur-containing materials, and isocyanates are derived
from polyurethanes.

The first stage in the thermal degradation of polyurethanes
involves depolymerisation to give the monomers (diisocyanates
and diols):   25

$$[-CONH(CH_2)_nNHCO_2(CH_2)_mO-]_p \rightarrow$$
$$pOCN(CH_2)_nNCO + pHO(CH_2)_mOH$$

Some of the diisocyanate is evolved as a trimer, forming a
characteristic yellow smoke. Further degradation occurs in
several ways; for example: 30

1. formation of carbodiimide and carbon dioxide:

$$2RNCO \rightarrow R-N=C=N-R + CO_2$$

2. the diol undergoes dehydration and charring, and evolves
   some low molecular weight hydrocarbons;

3. formation of hydrogen cyanides and nitriles such as pro-   35
panonitrile.

The nitriles will then degrade further. The pyrolysis of
propanonitrile has been studied in detail, and a complex free
radical mechanism has been proposed. The suggested initia-
tion step is the rupture of the weakest bond in the molecule: 40

$$C_2H_5CN \rightarrow CH_3 + \cdot CH_2CN$$

This is followed by a series of propagation steps, leading to the
formation of many products, including hydrogen, methane,
ethane, ethene, and hydrogen cyanide. Termination is brought
about by the combination of two of the more stable radicals. In   45
a fire, oxidation of these products may then occur, leading to
water and oxides of carbon and nitrogen, but no intermediate
oxidation products (such as aldehydes and ketones) have ever
been observed.

Although this mechanism fits most of the observed facts,   50
there are still some unexplained problems. The difficulty is
that the actual conditions found in a real fire will vary
enormously; such factors as the amount of oxygen available
and the temperature are important.

Adapted from an article with the same title by R. F. Simmons and E. Metcalf,
Chemistry in Britain, *March 1987*

1 Which is the weakest bond in the propanonitrile molecule
(line 40)?   [2]

2 Write the empirical formula of polyvinyl chloride, and an equation
for its complete combustion. Use this to calculate the mass
percentage of HCl in the product mixture, and compare it with the
figure given (lines 19–20). How do you account for the difference?
(C = 12, H = 1, O = 16, Cl = 35·5)   [8]

3 Draw displayed formulae for the following:
  a phosgene (line 4),
  b one unit of the polyurethane, in which $n = m = 1$ (line 26),
  c the diisocyanate in which $n = 1$ (line 26), and
  d propanonitrile (lines 35–6).   [8]

4 Explain carefully the term 'free radical' (lines 38–9). Why are most
mechanisms which involve free radicals complex?   [3]

5 Write *three* possible reactions for the oxidation of hydrogen cyanide, each involving different amounts of elemental oxygen, to account for *five* products (line 46). [6]

6 Write structural formulae for *one* aldehyde and *one* ketone which the investigators might have looked for here (line 48), as possible products of the intermediate oxidation. [2]

7 What kind of hybridisation does the central carbon atom in carbodiimide (lines 31–2) show? What shape will this lead to? Are any of the electrons delocalised? Justify your answer, with an appropriate diagram. [6]

Total = 35

Additional work

How can studies like the one outlined in the above passage (lines 23–49) help to prevent or contain fires where plastics are found? What other precautions might also help?

Exercise 3·5

Permanent future for biodegradables

biochemistry, environment, polymers, resources,
Third World

It is a conundrum: plastics for packaging have to be proof
against sunlight, air, damp, and micro-organisms, otherwise
they cannot do their job of containing and protecting the
products they enclose. Yet these very properties make them
difficult to dispose of—as our littered beaches and public places   5
clearly show. The development of plastics which are stable
enough for the uses they are put to, and yet are biodegradable
once they are discarded, is a challenge for the chemical
industry.

ICI has recently developed a biodegradable plastic called   10
Biopol. This is based on 3-hydroxybutanoic acid (HB), which is
polymerised by a certain bacterium, *E. eutrophus*, to give
polyhydroxybutanoate (PHB), a linear polyester, which the
bacteria use as a way of storing energy. The trick is to feed HB
to the micro-organism, together with a glucose fermentation   15
medium and another organic acid. The micro-organism then
produces a copolymer with the right characteristics, Biopol.

The beauty of Biopol from an environmental point of view is
that not only is it totally biodegradable by soil organisms to
carbon dioxide and water, but it is manufactured from renew-   20
able resources. It can be incinerated safely (it contains no
halogens), but its inherent instability will rule it out for some
long-life uses and for recycling. Also, since PHB is food for
bacteria, it may be unsuitable for food packaging. ICI claims
that in normal use Biopol is comparable in durability, stability,   25
and water resistance to conventional thermoplastics. Upon
burial in landfill or composting with sewage, however, it can
break down in a matter of weeks. After extensive trials, Biopol
is now being used by Wella, the hair-care company, even
though it costs much more than conventional thermoplastics.   30

Meanwhile, researchers around the world are trying to make
different forms of PHB. Since the PHB-synthesising gene was
identified in *E. eutrophus* in 1987, scientists have been transferring the gene to other micro-organisms ('genetic
engineering') in the hope of making PHB more cheaply or with   35
better properties. Even more exciting is the possibility of
putting the PHB gene into plants like potatoes or turnips, so

that they make PHB instead of starch. This would be an extremely convenient and cheap way to make biodegradable plastic.

Starch itself can be incorporated into conventional plastics such as polyethene, as much as 15%, to make a biodegradable product. When the plastic is discarded, micro-organisms eat the starch and the plastic is reduced to a powder. Once this happens, oxidation will reduce the molecular weight, and bacteria can then attack the polythene.

Some polymer manufacturers argue, however, that recycling is the answer to the problem of waste plastic, and that the presence of biodegradables would have a disastrous effect on this policy and thus should not be developed. Others believe that, although recycling of plastic sounds like a good idea, it can never become a practical reality on a significant scale.

Adapted from an article with the same title by R. Stevenson, Chemistry in Britain, *June 1990*

1 Draw a displayed formula of 3-hydroxybutanoic acid, and *three* units of the polyester derived from it (lines 11–12). [4]

2 What is the difference between a copolymer (line 17) and a homopolymer? [3]

3 What is meant by the term 'conventional thermoplastic' (line 26)? Give an example, its empirical formula, and one of its important uses. [5]

4 List the advantages and disadvantages of a biodegradable plastic such as Biopol. [6]

5 Write a balanced equation for the complete oxidation of HB (lines 19–20). [3]

6 Why should reducing the plastic to a powder facilitate oxidation (line 44)? [3]

7 What would you see as the advantages and disadvantages of turning a starch-producing plant into a plastic-producing plant (lines 36–8), particularly for the Third World? [6]

Total = 30

Additional work

What are your views on the opinions expressed in the last paragraph?

Wolves in sheep's clothing

chlorine, environment, free radical mechanism, ozone layer, photolysis

The chlorofluorocarbons (CFCs) developed in the 1930s are quite remarkable. They are excellent refrigerants, solvents, and carrier gases, are non-toxic and non-flammable, and are remarkably unreactive. This stability is their environmental drawback. Since they are transparent, they do not photolyse in 5 visible light; they do not dissolve in water; they do not oxidise. None of the usual environmental sinks is applicable, and hence they can be found in places remote from their point of manufacture or use (e.g., in Ireland and Antarctica).

However, as the compounds rise in the atmosphere they 10 become exposed to increasing UV radiation, and this does eventually cause photolysis of the C—Cl bonds.

The Earth's surface is protected from much of the sun's UV by the absorption of these wavelengths by ozone. A balance between UV-photolysis and reactions involving molecular 15 oxygen maintains a layer of increased ozone concentration some 25–35 km above sea level:

$$O_2 \overset{h\nu}{\to} 2O$$

followed by $\quad\quad O + O_2 \to O_3$

Unfortunately, if a CFC molecule gets involved in these 20 photolysis reactions, it destroys the equilibrium by setting off a chain reaction of ozone destruction; for example:

$$CFCl_3 \overset{h\nu}{\to} CFCl_2^{\cdot} + Cl\cdot$$

followed by $\quad\quad Cl\cdot + O_3 \to ClO\cdot + O_2$

and $\quad\quad ClO\cdot + O \to Cl\cdot + O_2$ 25

The chain length before Cl finally finds its way back into the lower atmosphere is about 100 000 reactions. One chlorine atom can destroy that number of ozone molecules. Rowland and Molina, who first reported this phenomenon in 1974, realised that the atmosphere has only a finite capacity to 30 absorb the Cl atoms produced in this way, that the amount of CFCs being released into the atmosphere probably exceeded this capacity, and that the ozone layer could be irreversibly depleted. This would lead to increased numbers of skin cancer cases (due to increased UV radiation reaching the Earth's 35 surface), and possible profound changes in climate.

This caused a stir in the USA, and the use of CFCs in aerosols (perceived to be a trivial use) was banned in both that country and Scandinavia. Western Europe and the rest of the world, however, were not convinced; where was the *evidence* that the ozone layer was being damaged? Even in the USA there was much criticism of the ban, and CFCs continued to be poured into the atmosphere. Politicians thought the problem had gone away, and CFCs were soon being made on as large a scale as before.

Then, quite unexpectedly, the smoking gun was found. The British Antarctic Survey had been keeping records of the Antarctic atmosphere since 1957, and in 1982 they noticed a progressive diminution in stratospheric ozone each summer. A hole was opening in the ozone layer over Antarctica. Since then there has been an increasing amount of scientific activity, and it is now known that related compounds containing bromine are up to ten times more damaging to the ozone layer, but that chlorohydrocarbons are much less so (since they are largely destroyed before they get as high as the ozone layer).

There has been much activity on the political front too, culminating in the Montreal Protocol of 1990, in which the participating nations agreed to phase out fully-halogenated compounds by the year 2000. However, the ban does not yet include the molecules containing some hydrogen, and nor has it been signed by many developing countries, including India and China, where the use of CFCs is increasing most rapidly.

Adapted from an article by R. Stevenson in Chemistry in Britain, *August 1990*

1 What are 'environmental sinks' (line 7)? [2]

2 a Explain the meaning of 'photolysis' (line 12).
  b Why is it that compounds which are transparent to light are not photolysed by it (lines 5–6)? [6]

3 What is the difference between light and UV radiation, and why do CFCs respond in distinctly different ways to them? [4]

4 Give the systematic name of the reactant in the equation given on line 23, and draw a displayed formula. Explain the fact that the F—C—Cl bond angle is larger than the Cl—C—Cl bond angle. [6]

5 The products of this same reaction are called free radicals. Explain briefly why they are such reactive species, and why they undergo chain reactions. How can such chain reactions terminate? [6]

6 What do you understand by the phrase 'the smoking gun' in this context (line 46)? [2]

7 Why do you think that many developing countries are as yet refusing to implement the phasing out of CFCs? What could be done about this? [4]

Total = 30

Additional work

Given that CFCs have many uses (only a few of which are mentioned in the article), for some of which there are no alternatives available, do you think that the phasing out is justified? Should they be banned immediately, due to the strong circumstantial evidence of their damaging effect on the ozone layer?

Exercise 3·7

The conversion of methane

*catalysis, energy, equilibrium, global warming,
industry, resources*

Natural gas (generally 90 mole percent or greater of methane)
is a very large resource, but methane is as yet very much
secondary to petroleum oil in providing for our energy and
petrochemical needs. This is for both economic and chemical
reasons, which can be summarised as follows: 5

1. Most large reserves are thousands of miles from where they
   could be used, often associated with liquid petroleum. Since
   natural gas has a low energy density, it is not economic to
   transport it long distances, and so it is flared off. This
   represents a considerable waste of energy and is a signi-   10
   ficant contributor to global warming.

2. Thermodynamically, methane is much more stable than
   other hydrocarbons, and is the only one which is stable with
   respect to its elements. Hence any attempt at pyrolysis to
   form higher hydrocarbons would be equilibrium controlled   15
   and result in negligible yields. In addition, the C—H bond is
   extremely strong (bond energy = 440 kJ mol^{-1}), far stronger than most other bonds, and methane is surprisingly
   stable, even at temperatures where its decomposition might
   be predicted on thermodynamic considerations alone.   20

   The rewards would be great if a practical way of converting
methane into methanol or higher hydrocarbons could be found;
such products could be used both as fuels and as chemical
feedstocks. Hence despite the formidable chemical problems, a
considerable research effort is being made, with at least some   25
success. Here, we will consider three recent attempts.

Direct oxidation
The partial oxidation of methane is energetically favourable,
and looks like a simple operation:

$$CH_4 + \tfrac{1}{2}O_2 \rightarrow CH_3OH \quad \Delta H = -126 \text{ kJ mol}^{-1}$$ 30

In practice, it is very difficult to prevent further oxidation of
the methanol. Very mild oxidising catalysts and a low proportion of oxygen have been tried with minimum success. The
main hope seems to be to find a way of removing the methanol

very rapidly after it has been formed, and surprisingly enough, 35
homogeneous catalysts look promising. The mechanism of the
reaction is fairly well understood, and seems to proceed by a
series of free radical steps, the initial one being:

$$CH_4 + O_2 \rightarrow CH_3^{\cdot} + HO_2^{\cdot}$$

Oxidative coupling
40
Methane can be made to react directly with oxygen, using an
appropriate catalyst, without any of the oxygen being retained
in the organic product:

$$2CH_4 + \tfrac{1}{2}O_2 \rightarrow C_2H_6 + H_2O$$
and
$$2CH_4 + O_2 \rightarrow C_2H_4 + 2H_2O$$
45

Alternatively, methane at high pressure and temperature is
passed over an easily reduced metal oxide (such as a manganese oxide). When the oxide has all been reduced, it is
oxidised back to its former state by passing oxygen over it.

Reactions involving chlorine
50
Methane will react with chlorine to form chloromethane which
will pyrolyse, at sufficiently high temperatures, to give ethene
and ethyne. A promising alternative is the oxyhydrochlorina-
tion of methane over a chloride catalyst at 350 °C. For example:

$$CH_4 + HCl + \tfrac{1}{2}O_2 \rightarrow CH_3Cl + H_2O$$
55

Other chloromethanes are also produced simultaneously, and
the mixture of these is passed over another catalyst at the
same temperature to produce a mixture of hydrocarbons which
constitutes a gasoline. The other product is HCl which is
recycled. The main problem with this process seems to be the 60
highly corrosive nature of the reactant and product mixtures.

Once the technical problems of these and other processes
have been solved, their commercial viability will very much
depend upon the price of crude oil; the more expensive this
becomes, the more attractive these alternative sources of fuel 65
and feedstock will be.

Adapted from an article by N. D. Parkyns in Chemistry in Britain,
September 1990

1 Assuming the other 10% to be ethane, what would be the average RMM of natural gas (line 1)? [3]

2 What do you understand by the term 'low energy density' (line 8), as it applies to natural gas? [3]

3 Briefly explain the meaning of the following terms, and give an example of each:
   a equilibrium controlled (line 15);
   b chemical feedstock (lines 23–4);
   c homogeneous catalysts (line 36). [9]

4 What other consideration might be controlling this reaction (line 20)? [2]

5 What products might result from further oxidation of methanol, and why should this be so difficult to prevent (lines 31–2)? [5]

6 Define 'pyrolyse' (line 52). For *either* of the two pyrolysis reactions mentioned here, give a balanced equation. [4]

7 Choosing suitable oxides of manganese, write equations to illustrate this reversible process (lines 46–9). [4]

Total = 30

Additional work

Explain ways in which the flaring of natural gas contributes to global warming (line 11).

Exercise 3·8

The Teflon story

fluorine, Hess's Law, % composition, polymers

In April 1938, a chemist working for Dupont in the USA, Roy
Plunkett, prepared the gas tetrafluoroethene by dechlorinating
$CClF_2CF_2Cl$ with zinc. He was going to react it with HCl under
pressure to make CHF_2CF_2Cl, which might be usable as a
refrigerant. However, when he came to release some of the   5
fluoroalkene from the cylinder in which he was storing it, none
came out. Thinking that the valve had leaked, he dismantled
it, and found a white powder inside the cylinder. Analysis
showed that the empirical formula was CF_2 and it was realised
that it was polytetrafluoroethene, PTFE, now known by the   10
Dupont trade name of Teflon. A lucky discovery, but Plunkett
was much luckier than he realised: it was later found by
experience that the monomer under pressure can explode
violently, giving a mixture of carbon and carbon tetrafluoride,
with half the energy release of the same mass of TNT—he was   15
lucky to be alive!

PTFE is a most unusual plastic: it is the most slippery
substance in the world (more slippery than wet ice!), and has
unique non-stick properties. It is an extremely efficient electrical insulator, and is renowned for its chemical inertness,   20
insolubility, weatherability and impermeability to moisture.
However, it is very difficult to work: it doesn't melt until about
340 °C, and due to the extremely high viscosity of the liquid it
cannot be fabricated by the normal methods of injection
moulding. It is also very expensive, and Dupont decided not to 25
invest any more in its development.

This changed, however, with the rush to build an atomic
bomb. In order to do this, the isotopes of uranium had to be
separated, and the best method was by the gas diffusion of
uranium hexafluoride, the only volatile compound of this heavy 30
metal. UF_6 and fluorine from which it is made are extremely
corrosive, attacking all the apparatus and lubricants then in
use. PTFE was found to be the only material which could resist
these two, and so its development was revived, under a tight
security net. After the war, peaceful uses were sought for   35
Teflon, and its unique properties found many indispensible
applications. For example, it is used for artificial hip joints and
tendons, and as sleeves for dough rollers in bakeries.

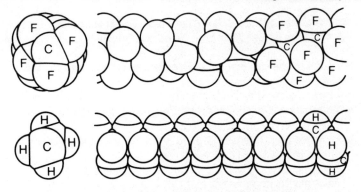

End and side views of the PTFE and polyethene chains.

It is interesting to compare PTFE with its hydrogen analogue, polyethene. Fluorine is larger than hydrogen—the van   40
der Waal's radii are 0·155 and 0·120 nm respectively, hence the
PTFE molecule is twisted (see diagram above). The C—F bond
is one of the strongest covalent single bonds, considerably
stronger than C—H, and hence not easily broken. Furthermore, the large F atoms effectively shield the much weaker   45
C—C bond from attack [E(C—F) = 467, E(C—H) = 413,
E(C—C) = 347 kJ mol^{-1}].

Adapted from Teflon Touches Gold, *R. E. Banks,* Chemistry in Britain,
May 1988

1 Write balanced equations for the two reactions described in
   lines 2–4.   [4]

2 From the empirical formula given in line 9, calculate the
   percentage composition. (C = 12, F = 19)   [3]

3 Which property/ies of Teflon described in the second paragraph
   make it particularly suitable for
   a artificial hip joints, and
   b sleeves for dough rollers (lines 37–8)?
   Give reasons.   [6]

4 Why can the isotopes of uranium not be separated by chemical
   means? Upon what principle does the method mentioned here
   (lines 29–30) depend?   [4]

5 What is meant by the van der Waal's radius of an atom (lines 40–41)? How does it differ from the covalent radius? [4]

6 Explain why the size of the F atom causes the PTFE chain to twist (line 42, and diagram). [2]

7 a Write a balanced equation for the explosive decomposition of PTFE (lines 13–14).

   b From the bond energies given in lines 46–7, plus the additional values $E(C\!=\!\!=\!C) = 612\,kJ\,mol^{-1}$ and $\Delta H_{at(graphite)} = 716\,kJ\,mol^{-1}$, select appropriate data to calculate the approximate enthalpy change of this reaction. Assume that the carbon formed is graphite. [6]

8 Consider the molecular data given in the diagrams and the last paragraph. How can this explain
   a the unreactivity of Teflon, and
   b its non-stick and slippery properties?
Justify your answers. [6]

Total = 35

Phenol from crude oil

*acidity in organic compounds, industry, resources,
structural formulae*

Phenol was discovered in 1834 when it was isolated from coal
tar. It was first named carbolic acid (a name still sometimes
used), but given its present name seven years later. It was
found to have strong germicidal properties, and phenol was the
first antiseptic used in hospitals and operating theatres. For 5
several decades this was its principal use, and coal tar provided sufficient quantities for this purpose. However, 2,4,6-
trinitrophenol (picric acid), a high explosive made from phenol,
was needed in large quantities during the First World War for
filling explosive shells, and the previous source of supply of 10
phenol was no longer adequate. New methods for its manufacture had to be developed.

The first such method was expensive, and unsuitable for
continuous operation, but it was the best then available.
Benzene was sulphonated, and the product fused with sodium 15
hydroxide:

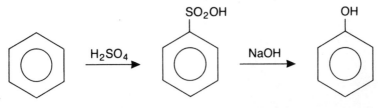

After the war, demand for phenol dropped to its prewar level,
and many countries were left with huge stockpiles. Soon after
this, however, phenol-formaldehyde resins (the first synthetic 20
thermosetting polymers) were developed, and demand again
increased. Stockpiles were rapidly used, and a new process for
the manufacture of phenol was introduced. Benzene was
chlorinated in the presence of iron(III) chloride, and the
chlorobenzene was heated with aqueous sodium hydroxide at 25
300 °C under pressure:

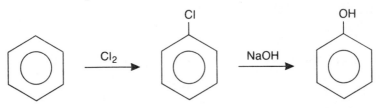

In a later development of this process benzene, HCl and air were passed over a catalyst at 200 °C, and the resulting chlorobenzene was reacted with steam at 425 °C over another catalyst: 30

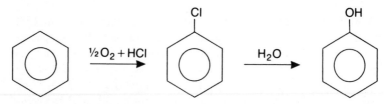

The other product, HCl, was recycled thus making the process more economically viable.

These methods all used benzene obtained directly from coal   35 tar as a starting material. The main source of organic compounds today, crude oil, does not contain benzene and so this must be made first. In the modern process, from which most of our phenol is obtained, suitable primary distillates are passed over a catalyst (either platinum or molybdenum(VI) oxide) in   40 the presence of excess hydrogen. Cyclisation and dehydrogenation occur; for example, hexane yields benzene:

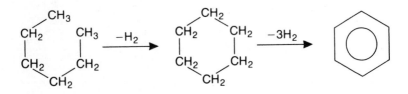

The benzene is separated from other products by solvent extraction and fractional distillation. It is then reacted with   45 propene to form (1-methylethyl)benzene using a suitable catalyst:

$$C_6H_6 + CH_3\!\!-\!\!CH\!\!=\!\!CH_2 \rightarrow C_6H_5CH(CH_3)_2$$

This product (often called cumene) is reacted with air to form the hydroperoxide, which yields phenol on treatment with   50 warm dilute sulphuric acid:

$$\begin{array}{c} CH_3 \\ | \\ C_6H_5CH(CH_3)_2 + O_2 \rightarrow C_6H_5\!\!-\!\!C\!\!-\!\!CH_3 \rightarrow C_6H_5OH + (CH_3)_2C\!\!=\!\!O \\ | \\ O\!\!-\!\!OH \end{array}$$

One of the advantages of this method of manufacture is that the other product is also an industrially important substance.

Over 80% of the phenol which is used today is made by this process. Within a very few decades, however, we shall have to find alternative routes for the manufacture of the aromatic starting material. 55

1 Write the structural formulae of
  a picric acid (line 8), and
  b (1-methylethyl)benzene (line 46). [4]

2 Explain the meaning of the terms
  a thermosetting (line 21);
  b continuous operation (line 14);
  c primary distillates (line 39);
  d solvent extraction (lines 44–5). [8]

3 Write a suitable equation and use it to help explain why picric acid is a high explosive (line 8). [5]

4 Explain in detail why phenol was called carbolic *acid* (line 2). Outline a series of simple experiments that you could perform to show the relative acidity of phenol compared to other organic hydroxyl compounds. [10]

5 In the two processes involving chlorobenzene, fairly extreme conditions must be used to replace the chlorine by the hydroxyl group. Explain why this is. [3]

Total = 30

Additional work

Why does the author suggest that we shall need to find alternative routes for the manufacture of aromatic compounds in the future? What alternatives may be available?

Exercise 3·10

Oils and fats go to market

environment, esters, nomenclature, structural
formulae, synthesis, world food problem

The world production of 17 major oils and fats (glycerides) over
a 40-year period is shown in the table. The growth is based
entirely on the four major ones—soybean, palm, sunflower, and
rape seed. The growth of the last two is partly because of the
desire of the European Community (EC) to become more   5
self-sufficient in vegetable oils, but there is a price to pay. The
cost of production per tonne in 1987 of the three major oils was
$215 for palm oil, $315 for soybean, and $750 for rape seed.

World production of major oils and fats (10^6 tonnes)

Year	1960	1970	1980	1990	2000 (est.)
Soybean	3·2	6·0	12·6	17·0	23·3
Palm	1·3	1·7	4·5	10·1	17·5
Sunflower	1·9	3·6	4·9	8·0	10·0
Rape seed	1·1	2·0	3·7	8·3	10·8
Other*	21·7	27·0	31·1	37·3	43·7
World production	29·2	40·3	56·8	80·7	105·3

* Cotton seed, groundnut, sesame seed, corn, olive, palm kernel, coconut, butter, lard,
tallow, fish, linseed, and castor.

About 80% of the oils and fats now produced is for food use,
another 6% of poorer grade is used in animal feed. The   10
remaining 14% is used by the oleochemical industry, mainly to
make surfactive materials such as traditional soaps. The world
annual average consumption per person in 1985 was 13·5 kg;
in the USA it was 38·5 kg, in the EC 36·2 kg, in China 6·5 kg,
and in India 7·0 kg. It would take a lot more production to raise   15
the last two countries up to the world average. The developed
world is not expected to decrease its consumption, because
even if we decrease our dietary requirements we shall use
more for surfactant products, as these are more environmentally friendly in terms of biodegradability than similar pro-   20
ducts based on petroleum oil.

Not only is world production increasing, but chemists,
biotechnologists and agriculturists are meeting consumer demands by rearranging the composition of existing oils, by

breeding new varieties of oil-bearing plants, and by using   25
so-far under-exploited plants. Chemical processes include:

• **Blending** oils from different sources.

• **Fractional crystallisation**, to separate the higher melting
  (more saturated) from the lower melting (more unsaturated)
  glycerides.   30

• **Partial hydrogenation**, to decrease the degree of un-
  saturation.

• **Interestification**, to rearrange the fatty acids among the
  glycerol units.

• Large-scale **chromatography** to separate specific glycer-   35
  ides in a pure state from natural oils (which are always
  mixtures).

The role of biotechnology is typified by the case of rape seed.
Erucic acid (structure (1)), which occurs in rape seed oil, has
been found to build up in the hearts of rats fed on it. Although   40
there is no evidence of a similar effect in humans, it has been
thought prudent to breed new varieties of this plant which are
low in this acid, for edible use. However, the amide of erucic
acid is used in plastic manufacture, so the older varieties are
still grown. Yet other varieties of rape are being developed   45
which give oils similar in composition to coconut oil, the price
of which fluctuates wildly due to political instability and
unreliable weather.

Agriculturists are developing ways of cultivating wild
plants, such as *Cuphea*, which is rich in the medium chain   50
acids (C_6–C_{14}), and coriander, which is rich in petroselinic acid,
an isomer of oleic acid (2), which can be oxidised to lauric (3)
and adipic (4) acids.

A variety of dietary and health claims have been made in
respect of certain polyunsaturated oils; for example, they are   55
thought to lower cholesterol levels by an as-yet undetermined
mechanism. These include γ-linolenic acid (6,9,12-octa-
decatrienoic acid) (5), EPA (5,8,11,14,17-eicosopentaenoic acid)
and DHA (4,7,10,13,16,19-docosahexaenoic acid). The last two
occur in fish oils, and can be obtained in capsules in health food   60
stores as glycerides or as ethyl esters. γ-linolenic acid occurs in
oil from the evening primrose, blackcurrant seeds, and also
from an oil produced by microbiological techniques.

$$CH_3(CH_2)_7CH{=}CH(CH_2)_{11}COOH$$
1. Erucic (*cis*-13-docosenic) acid

$$CH_3(CH_2)_7CH{=}CH(CH_2)_7COOH$$
2. Oleic (*cis*-9-octadecenoic) acid

$$CH_3(CH_2)_{10}COOH$$
3. Lauric (dodecanoic) acid

$$HOOC(CH_2)_4COOH$$
4. Adipic acid

$$CH_3(CH_2)_4CH{=}CHCH_2CH{=}CHCH_2CH{=}CH(CH_2)_4COOH$$
5. γ-linolenic (6,9,12-octadecatrienoic) acid

Adapted from an article with the same title by F. D. Gunstone,
Chemistry in Britain, *June 1990*

1 a Which of the four major oils has shown the greatest percentage increase in production in the period 1960–1990? Show your working and suggest a reason.

   b Verify the statement made in the second sentence (lines 2–4) by expressing the total production of the four major oils as a percentage of world production in 1960 and in 2000.　　　　　[5]

2 Explain briefly the following terms, as they are used here:
   a microbiological techniques (lines 63–4);
   b environmentally friendly (lines 19–20).　　　　　[4]

3 Using R to represent the hydrocarbon chain, outline a possible reaction scheme for the preparation of the amide from erucic acid (line 43).　　　　　[4]

4 By what process is partial hydrogenation (line 31) carried out, and what physical effect does this have on the glycerides?　　　　　[4]

5 Assuming that the oxidation occurs at the double bond in petroselinic acid, write its structural formula (lines 51–3), in the same way as the given formulae. What is the *systematic* name of adipic acid?　　　　　[5]

6 How would you make the ethyl ester of DHA from the glyceride (line 59)? Outline the essential chemical steps.　　　　　[5]

7 Compare the *structure* with the *systematic name* of γ-linolenic acid. Use this information to help you to write the structural formula of EPA (lines 58–9) [note: eicoso = 20].   [3]

Total = 30

Additional work

Suggest ways by which the average annual consumption of glycerides in India and China might be increased, to improve the standard of nutrition in these countries (lines 12–16).

ENERGY, EQUILIBRIUM AND KINETICS

The invention of dynamite

combustion, mechanism, thermochemistry

An explosion is the result of a large and rapid increase in volume. Although the largest explosion in recorded history—that of the volcano Krakatoa in 1883—was caused by seawater coming into contact with molten rock and vaporising instantly, most explosions are a result of chemical reactions. Essentially, 5 a reaction will have the potential to be explosive if it is highly exothermic and a high proportion of the products are gases. It seems ironic that all commercial explosives contain compounds of nitrogen, that most inoffensive and inert of elements. Alone amongst the elements found in explosives, it is released in 10 elemental form rather than as an oxide during the course of the explosion.

Whilst gunpowder, the original explosive, was a fortuitous discovery by alchemists, the new explosives of the 19th century were deliberately developed by industrial chemists to support 15 the advancing technologies of mining and civil engineering as well as military use. Traditional gunpowder ('black powder') consisted of saltpetre (potassium nitrate), charcoal and sulphur, with variations in the ratio and physical form to suit the circumstances. 20

The first modern explosive was nitroglycerine (propane-1,2,3-triol trinitrate), invented by professor Ascanio Sobrero of Turin in 1846. He recognised its possibilities, but the dangers of handling it frightened him. For some years, its only use was in the treatment of *angina pectoris*. Nitroglycerine, a heavy 25 oily liquid, was originally produced by pouring purified glycerine into a cooled nitrating mixture of nitric and sulphuric acids.

Black powder is a 'low' explosive—its power depends on the intermolecular reaction between its components, on its physi- 30 cal form, and on the environment. Nitroglycerine was the first 'high' explosive—acting by intramolecular reaction that is independent of other constraints. Furthermore, the reaction can be initiated by shock, hence its danger. Railway companies quickly banned its carriage when it was realised that simply 35 dropping a can of the capricious explosive might detonate it!

A young Swedish inventor, Alfred Nobel, was not deterred by nitroglycerine's fearsome reputation. He improved the yield of Sobrero's process by mixing the glycerine and sulphuric acid before adding the nitric acid. He built a plant to manufacture 40 it, and he doggedly marketed it to the mining industry as 'blasting oil', despite the opposition of various authorities and the death of his younger brother in a factory explosion. Later, he tamed nitroglycerine by absorbing it into natural siliceous materials, which did not affect its explosive powers, but 45 rendered it insensitive to shock. The result, dynamite, made Nobel's fortune.

Adapted from an article by R. Stevenson in Chemistry in Britain,
October 1989

1 From the information given in the first paragraph and lines 17–19, name *four* possible products of the explosion of gunpowder, and write a balanced equation to justify your choice. [6]

2 Explain why only nitrogen is not released as an oxide during a chemical explosion (lines 9–12), by considering the following bond energies (given in kJ mol^{-1}):

$$E(N{\equiv}N) = 945{\cdot}4 \quad E(N{-}O) = 214 \quad E(N{=}O) = 587 \qquad [4]$$

3 Write the displayed formula of propane-1,2,3-triol trinitrate given that its empirical formula is $C_3H_5N_3O_9$ (lines 21–2). [2]

4 a Write an equation for the explosion of nitroglycerine, again bearing in mind the information given in lines 10–11.
   b Compare this with the equation you have written in answer 1, and use these two equations to explain the difference between 'intermolecular' (line 30) and 'intramolecular' (line 32) in this context. [6]

5 What is the name of the key ion essential for the nitration of an organic compound? Write an equation to show how it is produced from nitric and sulphuric acids (lines 27–8).
   How would you classify this reaction? [4]

6 Write a possible mechanism for the reaction for the nitration of
nitroglycerine (lines 25–8). [5]

7 Can you suggest how the 'physical form' and 'the environment'
might affect the explosive power of black powder (lines 29–31)? [3]

Total = 30

The fight against air pollution

adsorption, bonding, catalysis, combustion, energy,
d-block, environment, shapes of molecules

Heterogeneous catalysis plays a major role in the fight against
urban air pollution. Two components of automobile exhausts
that are involved in the formation of photochemical smog are
unburnt hydrocarbons and nitrogen oxides. In addition, there
may be considerable quantities of carbon monoxide. Even with   5
the most careful attention to engine design and fuel characteristics, it is not possible under normal driving conditions to
reduce the contents of these pollutants to an acceptable level in
the gases coming from the engine. It is therefore necessary to
remove them from the exhaust gases before they are vented to   10
the air. This removal is accomplished by a *catalytic converter*
(see diagram).

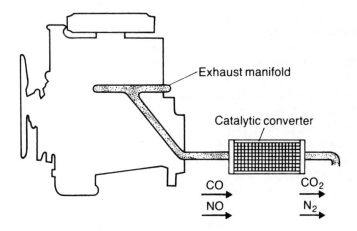

The exhaust system of an automobile, showing the position of the
catalytic converter.

The catalytic converter must perform two distinct functions:
(1) oxidation of unburnt CO and hydrocarbons to carbon
dioxide and water, and (2) the reduction of nitrogen oxides to   15
nitrogen gas. These two functions require two distinctly different catalysts. They must be effective over a wide range of
operating temperatures; they must continue to be active in

spite of the poisoning action of various petrol additives; they must be rugged enough to withstand gas turbulence and the 20 mechanical shocks of driving for thousands of miles.

Catalysts that promote the combustion of CO and hydrocarbons are, in general, the transition metal oxides and noble metals such as platinum. As an example, a mixture of two different metal oxides such as CuO and Cr_2O_3 might be used. 25 These materials are supported on a structure which allows the best possible contact between the flowing exhaust gas and the catalyst surface. Either bead or honeycomb structures made from alumina, Al_2O_3, and impregnated with the catalyst may be employed. Such catalysts operate by first adsorbing oxygen 30 gas, also present in the exhaust gas. This adsorption weakens the $O\!\!=\!\!O$ bond, so that oxygen atoms are effectively available for reaction with adsorbed CO to form CO_2. Hydrocarbon oxidation probably proceeds somewhat similarly, with the hydrocarbons first being adsorbed by rupture of a C—H bond. 35

Reduction of nitrogen oxides (for example, the decomposition of NO into N_2 and O_2) is favoured thermodynamically, but the reaction is extremely slow. A catalyst is therefore necessary. The most effective catalysts are the same kind as those for (1), but those which are most effective in (1) are usually less 40 effective in (2), and *vice versa*. It is therefore necessary to have two different catalytic components.

The activity of catalytic converters decreases with use, because of the loss of active catalysts, cracking and fractures due to repeated heating and cooling, and poisoning of the 45 catalysts. One of the most active catalyst poisons is the lead that comes from the tetramethyl lead, $Pb(CH_3)_4$, or tetraethyl lead, $Pb(C_2H_5)_4$, added to the gasoline. Because of severe catalyst poisoning that results from the use of leaded fuels, cars built since 1975 (in the USA) have been engineered to 50 discourage the use of leaded fuel.

Adapted from Chemistry, The Central Science, *T. L. Brown and H. E. LeMay,*
Prentice Hall (1988)

1 Explain the terms
  a heterogeneous (line 1);
  b photochemical (line 3). [4]

2 Give
  a the name and formula of a hydrocarbon with six carbon atoms which might be found in gasoline (petrol). Write an equation for the complete combustion of this hydrocarbon.
  b the formulae of *two* oxides of nitrogen which might be found in exhaust gases. How do you account for the presence of these oxides? [8]

3 Why do you think it is not possible to reduce the level of pollutants in the gases leaving a car engine (lines 7–9)? [3]

4 What property of transition metals enables them to catalyse reactions of the type discussed in the passage? [8]

5 Write a balanced equation for the decomposition of NO (lines 36–7), and use it to help you explain
  a the meaning of 'favoured thermodynamically' (line 37), and
  b why the reaction is extremely slow, and how a catalyst speeds it up (line 38); illustrate your answer with a sketch graph. [7]

6 What type of bonding, and what shape, would you predict for tetramethyl lead (line 47)? Briefly explain your reasoning. [6]

7 For what other reason, apart from being a catalyst poison (lines 46–8), is lead undesirable in exhast gases? [2]

Total = 38

Exercise 4·3
The bomb calorimeter

combustion, enthalpy, internal energy,
thermochemical calculations

If one were to ask which type of chemical reaction was the most important to mankind, then the answer would undoubtedly be the combustion of organic substances. For example, the burning of methane:

$$CH_{4(g)} + 2O_{2(g)} \rightarrow CO_{2(g)} + 2H_2O_{(l)}$$ 5

The precise study of reactions where complete combustion occurs is thus of great scientific interest and economic significance, and accurate ways of measuring the heat evolved during such reactions have been developed. To the chemist, the **standard enthalpy of combustion** ($\Delta H^{\ominus}_{combustion}$) of the 10 substance is the normal way of presenting the data, whereas the **fuel value** is more useful in industry—that is, the heat given out when 1 g of the substance is burnt.

The bomb calorimeter.

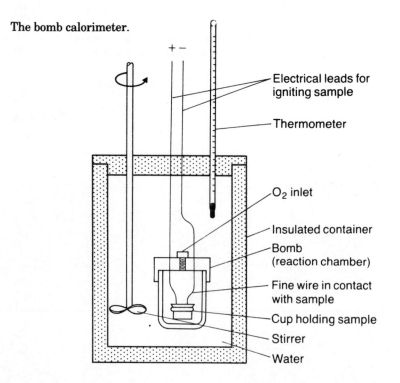

Combustion reactions are most conveniently studied in a **bomb calorimeter**, a device shown schematically in the figure. A weighed quantity of the substance to be studied (a liquid or solid) is contained in a small cup, within a vessel called a bomb. The bomb, which is designed to withstand high pressures, has an inlet valve for adding oxygen and also has electrical contacts to initiate the combustion reaction. It is filled with oxygen under pressure, sealed, and placed within the calorimeter. The calorimeter itself is a well insulated vessel containing an accurately measured quantity of water.

When all the components within the calorimeter have come to the same temperature, the combustion is initiated by passing an electric current through a wire in contact with the sample, which ignites when the wire gets hot. The heat evolved during the combustion is absorbed by the contents of the calorimeter, and when combustion is complete a steady temperature will again be reached after a short time interval. The temperatures before and after the combustion are carefully measured.

In order to use this information, it is necessary to know the specific heat capacity of the calorimeter: this is found by combusting a sample that gives off a known quantity of heat. For example, the combustion of exactly 1 g of benzoic acid, $C_7H_6O_2$, produces 26·38 kJ of heat. Suppose that 1 g of benzoic acid is combusted in our calorimeter, causing a temperature increase of 5·022 °C. The specific heat capacity, C, is given by:

$$\frac{\text{heat produced}}{\text{temperature rise}} = \frac{26 \cdot 38}{5 \cdot 022} = 5 \cdot 253 \text{ kJ } °C^{-1}$$

Once we know this figure, we can measure the temperature changes, ΔT, produced by other reactions and calculate the heat evolved, q. It is then a simple matter to convert this to the heat evolved for the combustion of one mole of the substance.

The values for the heats of combustion thus produced will be at constant *pressure*, and are known as **internal energy changes**, ΔU, whereas enthalpy changes, ΔH, relate to reactions at constant *volume*. The difference is small, but an easily computed correction factor must be introduced in order to calculate the standard heats of combustion of substances accurately.

In part adapted from Chemistry, The Central Science, *T. L. Brown and H. E. LeMay, Prentice Hall (1988)*

1 What is meant by 'complete combustion' (line 2)?   [3]

2 Define the standard enthalpy of combustion (line 10).   [4]

3 What small inaccuracy would be introduced into the calculated value of ΔU due to the method of ignition (lines 25–7), and how might this be allowed for?   [3]

4 Use the fuel value of benzoic acid (line 37) to calculate the value of $\Delta H^{\ominus}_{combustion}$ for this compound (ignore the correction outlined in the last paragraph): show your working. (H = 1, O = 16)   [3]

5 Derive an expression relating q, C, and ΔT (lines 39–44).   [2]

6 A 0·5865 g sample of lactic acid, $C_3H_6O_3$, is burnt in a bomb calorimeter whose specific heat capacity is 4·812 kJ $°C^{-1}$. The temperature rose from 23·10 °C to 24·95 °C. Calculate the fuel value and the standard enthalpy of combustion of lactic acid (again, ignore the $\Delta U \rightarrow \Delta H$ correction).   [11]

7 How might the procedure outlined in the second paragraph (lines 14–23) be modified to measure the heat of combustion of methane (line 5)?   [5]

Total = 30

Additional work

Would you agree with the view put forward in the first sentence? If your answer is negative, propose an alternative. In either case, justify your answer.

Fuel cells

electrode potentials, emf, industry, power generation

Fuel cells are essentially devices for producing electricity
directly from combustion. They comprise two electrodes
(cathode and anode) and an electrolyte. The fuel, hydrogen, is
fed to the anode, and air or oxygen is fed to the cathode (Figure
1). The electrodes have several functions: they catalyse the   5
oxidation of the fuel (anode) and the reduction of the oxygen
(cathode), they function as interfaces between the reactants
and the electrolyte, and they conduct electrons to the current
collector or reaction site. The electrolyte must conduct ions
(but not electrons, to prevent a short circuit), and serve as a   10
separator to prevent the fuel and oxidant reacting directly
inside the cell. It may be supported by an inert matrix.
Electrochemical reactions occur at the electrode–electrolyte
interfaces, by which chemical energy is converted directly into
electrical energy. Some heat is also generated.   15

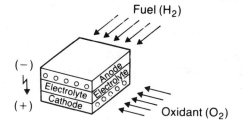

Figure 1   A schematic drawing of a fuel cell, showing basic
structure.

Many types of fuel cell are currently being developed for
large-scale use; two of the most promising are outlined below.

1. The phosphoric acid fuel cell (PAFC), which operates at
   150–200 °C. The electrolyte is 95% phosphoric acid in a
   silicon carbide matrix, and the electrodes are carbon-   20
   supported platinum. The electrode half reactions are:

$$\tfrac{1}{2}O_2 + 2H^+ + 2e^- \to H_2O \quad \text{and} \quad H_2 \to 2H^+ + 2e^-$$

   This is at present the furthest developed type of fuel cell,
   and it should be commercially available within a decade.

2. The alkaline fuel cell, which operates at 60–90 °C. The   25
electrolyte is 30–75% potassium hydroxide, and a variety of
electrodes have been investigated; amongst them nickel,
silver, and platinum. The electrode half reactions are

$$\tfrac{1}{2}O_2 + H_2O + 2e^- \rightarrow 2OH^- \quad \text{and} \quad H_2 + 2OH^- \rightarrow 2H_2O + 2e^-$$

A serious disadvantage of this cell is that carbon dioxide   30
must be removed from the air before it can be used as the
oxidant.

The standard emf of these cells is 1·32 V, but in practice
voltages of 0·7–1·0 V are obtained with current densities of
100–300 mA cm^{-2} of electrode surface. A large-scale fuel cell   35
plant will usually consist of three parts: a reformer to convert
the primary fuel into a hydrogen-rich gas, a stack of fuel cells
connected in series, and a device to convert the DC current to
AC (Figure 2).

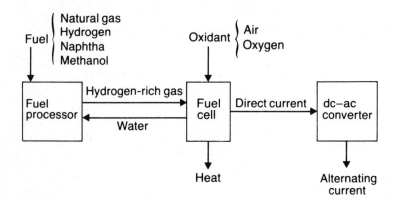

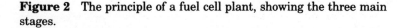

Figure 2 The principle of a fuel cell plant, showing the three main
stages.

Fuel cells are clean, silent and flexible. They produce no toxic   40
or corrosive products, and because they contain no moving
parts they are easily maintained. These advantages are at
present outweighed by the main disadvantages for industrial
power generation—they are very expensive, since the
materials used in their manufacture must meet high specifica-   45
tions, and their efficiency is limited by the need to reform the
primary fuel and the conversion to AC. They are, however,
widely used as portable power units where hydrogen can be

utilised directly as a fuel, where DC current is needed, and where cost is not of primary importance, such as in space 50 probes and in military applications. In the distant future, research and development may enable large fuel cell plants to be sited next to coal gasification plants to produce 150– 1000 MW of power. This would be a much more efficient way of generating electricity than the conventional or nuclear power 55 station.

Adapted from Fuel Cells Revisited, *G. J. Kleywegt and W. L. Driessen,*
Chemistry in Britain, *May 1988*

1 Consider the half reactions on line 22. Which one occurs at the cathode? Which ion is conducted through the electrolyte? [3]

2 Write the equations for the overall reactions occurring in the two sets of half reactions given on lines 22 and 29 respectively. [3]

3 Where might the heat come from to maintain the two cells discussed on lines 18–32 at their working temperatures? [3]

4 Making an appropriate approximation, calculate the concentration of 30% potassium hydroxide (line 26), in mol dm^{-3}. Show your working. (K = 39, O = 16, H = 1) [5]

5 With which component of the alkaline fuel cell would carbon dioxide (line 30) react? Give an equation. [3]

6 Explain what is meant by 'standard emf' (line 33), using the PAFC cell to illustrate your answer. Why is the working voltage of a fuel cell (line 34) not the emf? [5]

7 Name a readily available gas that might be used as a primary fuel (line 37). What is the essential process occurring in its reformation to hydrogen? [3]

Total = 25

Additional work

Outline the essential energy-conversion steps in the production of electricity from combustion in conventional power stations (line 55), and give reasons why they are less efficient than fuel cells.

Exercise 4.5

The buffering action of blood

biochemistry, combustion, displayed formulae, equilibrium, isomerism, Le Chatelier's principle, pH/buffers

Blood is an important example of a buffered solution. Human blood is slightly basic, with a pH of between about 7·39 and 7·45. In a healthy person, the pH never departs more than perhaps 0·2 pH units from the average value. Whenever the pH falls below about 7·4, the condition is known as *acidosis*; 5 whenever it rises above 7·4 it is known as *alkalosis*. Death may result if the pH falls below 6·8 or rises above 7·8. Acidosis is the more common tendency, because ordinary metabolism produces several acids.

The body uses three primary methods to control blood pH: (1) 10 the blood contains several buffers, including H_2CO_3–HCO_3^- and $H_2PO_4^-$–HPO_4^{2-} pairs, and haemoglobin-containing conjugate acid-base pairs. (2) The kidneys serve to absorb or release $H_{(aq)}^+$. The pH of urine is normally about 5·0 to 7·0. Acidosis is normally accompanied by increased loss of body fluids as the 15 kidneys work to reduce $H_{(aq)}^+$. (3) The concentration of $H_{(aq)}^+$ is also altered by the rate at which CO_2 is removed from the lungs. The pertinent equilibrium is:

$$H_2CO_{3(aq)} \rightleftharpoons H_2O_{(aq)} + CO_{2(aq)}$$

Removal of CO_2 shifts this equilibrium to the right, thereby 20 reducing $[H^+_{(aq)}]$.

Acidosis or alkalosis disrupts the mechanism by which haemoglobin transports oxygen in the blood. Haemoglobin (Hb) is involved in a series of equilibria whose overall result is

$$HbH_{(aq)}^+ + O_{2(aq)} \rightleftharpoons HbO_{2(aq)} + H_{(aq)}^+ \qquad 25$$

In acidosis, this equilibrium is shifted to the left, and the ability of haemoglobin to form oxyhaemoglobin (HbO_2) is decreased. The lesser amount of O_2 thereby available to cells in the body causes fatigue and headaches; if great enough, it also triggers 'air hunger' (the feeling of being out of breath that 30 causes deep breathing).

Temporary acidosis occurs during strenuous exercise, when energy demands exceed the oxygen available for complete oxidation of glucose to CO_2. In this case the glucose is converted to an acidic metabolite, lactic acid, $CH_3CHOHCOOH$. 35

Acidosis also occurs when glucose is unavailable to the cells. This situation can arise, for example, during starvation or as a result of diabetes. In the case of diabetes, glucose is unable to enter the cells because of inadequate insulin, the substance responsible for passage of glucose from the bloodstream to the interior of cells. When glucose is unavailable, the body relies for energy on stored fats, which produce acid metabolism products.

40

Adapted from Chemistry, The Central Science, *T. L. Brown and H. E. LeMay, Prentice Hall (1988)*

1 Define the term 'pH' (line 2). [1]

2 What is meant by a 'conjugate acid-base pair'? Give an example taken from the text (lines 12–13). [5]

3 What is meant by a buffer? Use the first of the two buffers given in lines 11–12 to explain how a buffer works, giving the appropriate equilibrium and equilibrium constant. [9]

4 Use the equilibrium given on line 19, and the one you have written in question 3, to explain how (3) works (lines 16–21). [4]

5 Use Le Chatelier's principle to explain why the equilibrium given on line 25 is shifted to the left in acidosis. [3]

6 Write an equation for the complete oxidation of glucose (lines 33–4).
Use it to explain why an insufficient supply of oxygen leads to acidosis. [4]

7 Draw the displayed formula of lactic acid (line 35). What type of isomerism would this compound display?
Using a suitable convention, draw the isomers. [6]

8 To what class of chemical compounds do fats belong? Draw the structure of a typical fat, using R to represent alkyl groups (line 42). [3]

Total = 35

Hydrogen peroxide and hydrazine

acid–base reactions, catalysis, displayed formulae,
electrode potentials and cells, Hess's Law calculation,
hydrogen bonding, K_b, safety

Hydrogen peroxide, H_2O_2, contains an O—O bond, and is the most familiar and commercially important peroxide. When pure, it is a colourless, syrupy liquid, density $1·47 \text{ g cm}^{-3}$, mp $-0·4 \,°C$, bp $151 \,°C$; these properties are characteristic of a highly polar, hydrogen-bonded substance. The pure liquid and 5 concentrated solutions are dangerously reactive, since they decompose with explosive violence if they come into contact with one of the many substances (see below) which catalyse the reaction:

$$2H_2O_{2(l)} \to 2H_2O_{(l)} + O_{2(g)} \quad \Delta H^\Theta = -196 \text{ kJ mol}^{-1}$$ 10

Hydrogen peroxide is marketed as a chemical reagent in aqueous solution of up to 30% by weight. A 3% solution is sold as a mild antiseptic; somewhat more concentrated solutions are employed to bleach fabrics such as cotton, wool, and silk. The peroxide ion is a by-product of metabolism that results 15 from the reduction of molecular oxygen, O_2. The body disposes of this reactive species with enzymes such as peroxidase and catalase: this accounts for the fizzing that occurs when dilute hydrogen peroxide is applied to an open wound, as the enzymes decompose it according to the above equation. 20

Hydrogen peroxide is capable of acting as an oxidising or reducing agent, as the two half equations show:

$$2H^+_{(aq)} + H_2O_{2(aq)} + 2e^- \to 2H_2O_{(l)} \quad E^\Theta = 1·77 \text{ V}$$

$$H_2O_{2(aq)} \to O_{2(g)} + 2H^+_{(aq)} + 2e^- \quad E^\Theta = -0·77 \text{ V}$$

It will, for example, oxidise I^- to I_2 and reduce the man- 25 ganate(VII) ion to Mn(II). Furthermore, the E^Θ values show that hydrogen peroxide is capable of disporportionating, but since this reaction is controlled kinetically it will only occur at a significant rate in the presence of a suitable catalyst, such as the enzymes mentioned above, or a number of inorganic 30 species, for example MnO_2 or Br_2:

$$[MnO_{2(s)} + 4H^+_{(aq)}],[Mn^{2+}_{(aq)} + 2H_2O_{(l)}] \,|\, Pt \quad E^\Theta = 1·09 \text{ V}$$

$$Br_{2(aq)}, 2Br^-_{(aq)} \,|\, Pt \quad E^\Theta = 1·23 \text{ V}$$

Hydrazine, N_2H_4, contains an N—N bond, and hence bears the same relationship to ammonia as hydrogen peroxide does to water. It may be prepared by the reaction of ammonia with the chlorate(I) ion: 35

$$2NH_{3(aq)} + OCl^-_{(aq)} \rightarrow N_2H_{4(aq)} + Cl^-_{(aq)} + H_2O_{(l)}$$

The poisonous nature of hydrazine, and the possibility of this reaction occurring, partly accounts for the oft-repeated warning against mixing household cleaners, many of which contain either ammonia or bleach (sodium chlorate(I) solution). 40

The physical properties of hydrazine are quite similar to those of hydrogen peroxide: it is a colourless, oily liquid with an mp of 1·5 °C and bp of 113 °C. It explodes on heating, and is a powerful reducing agent. The combustion of hydrazine is highly exothermic: 45

$$N_2H_{4(l)} + O_{2(g)} \rightarrow N_{2(g)} + 2H_2O_{(l)} \quad \Delta H^\ominus = -534 \text{ kJ mol}^{-1}$$

and hence it and one of its derivatives, methyl hydrazine, are used as rocket fuels. 50

An aqueous solution of hydrazine is weakly basic:

$$N_2H_{4(aq)} + H_2O_{(l)} \rightleftharpoons N_2H^+_{5(aq)} + OH^-_{(aq)} \quad K_b = 1·3 \times 10^{-6}$$

and salts containing the hydrazinium ion can be formed.

Adapted from Chemistry, The Central Science, *T. L. Brown and H. E. LeMay,*
Prentice Hall (1988)

1 Draw displayed formulae for
 a hydrogen peroxide (line 1),
 b methyl hydrazine (line 49), and
 c the hydrazinium ion (line 53). [3]

2 Outline how you would prepare a sample of crystalline
 hydrazinium sulphate, with the relevant equation. What safety
 precautions would you take, and why? [8]

3 a By drawing a suitable diagram, indicate the nature of the
 hydrogen bonding in hydrogen peroxide (line 5).
 b Would you expect hydrogen bonding to occur in hydrazine?
 Justify your answer, and cite any evidence in the passage which
 would support your view. [6]

4 From the equations given in lines 10 and 48, and the additional information that the standard heat of formation of water is $-285 \cdot 8$ kJ mol^{-1}, calculate the standard heats of formation of hydrogen peroxide and hydrazine. Draw suitable Hess's triangles to show your method. [11]

5 Can you suggest one reason why the reaction given on line 48 is so exothermic? [1]

6 What is meant by 'controlled kinetically' (line 28)? Write suitable equations to show how bromine catalyses the decomposition of hydrogen peroxide. [5]

7 Write a cell diagram for the cell made from the two electrodes given in lines 32 and 33, and calculate its emf. Give a balanced equation for the reaction which would occur if the cell were short-circuited, indicating clearly the *direction* of reaction. [7]

8 Write an expression giving K_b for this reaction (line 52). What name is given to this constant, and how does it show that hydrazine is weakly basic? [4]

Total = 45

Additional work

Write balanced equations to show how manganese(IV) oxide catalyses the decomposition of hydrogen peroxide (lines 21–33). How do the $E^\ominus$ values enable you to predict the *possible* catalytic action of this oxide, and what other factor(s) must be considered?

The Lewis theory of acids and bases

acid–base reactions, bonding, hydration, Lewis structures

For a substance to be a proton acceptor (that is, a base in the Brønsted–Lowry sense), that substance must possess an unshared pair of electrons for binding to the proton; as in, for example, the ammonia molecule:

$$H^+ + :NH_3 \rightarrow NH_4^+$$ 5

G. N. Lewis, an American chemist, was the first to notice this aspect of acid-base reactions. He proposed new definitions of acids and bases which emphasised the shared electron pair: a **Lewis acid** is defined as an electron-pair acceptor, and a **Lewis base** as an electron-pair donor. Everything that is a 10 base in the Brønsted–Lowry sense (i.e., a proton acceptor) is also a base in the Lewis sense (an electron-pair donor). However, the Lewis definition of an acid is much broader than the Brønsted–Lowry definition, since many more species than protons can accept the electron pair and hence can be con- 15 sidered as acids; for example, the reaction between ammonia (a base according to both definitions) and boron trifluoride. Boron has a vacant orbital in its outer shell, and therefore acts as an electron-pair acceptor (a Lewis acid):

$$H_3N: + BF_3 \rightarrow H_3N—BF_3$$ 20

The Brønsted–Lowry definitions are still regarded as the most useful, since when we think of acid–base reactions we are normally considering proton transfers in aqueous solution. The advantage of the Lewis theory is that it allows us to treat a wider variety of reactions as acid–base reactions, including 25 many which do not involve protons. Any compounds in which there is an incomplete octet of electrons, such as BF_3, are Lewis acids, and many simple cations (particularly those of the transition metals) can function as Lewis acids. For example, the Fe^{3+} ion interacts strongly with cyanide ions to form the 30 hexacyanoferrate(III) ion:

$$Fe^{3+}_{(aq)} + 6CN^-_{(aq)} \rightarrow [Fe(CN)_6]^{3-}_{(aq)}$$

Some compounds with multiple bonds can behave as Lewis acids. For example, the reaction of carbon dioxide with water to

form carbonic acid can be pictured as an attack by a water   35
molecule on the carbon atom of carbon dioxide:

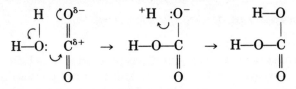

A similar kind of Lewis acid–base reaction takes place when
any non-metal oxide dissolves in water.

The Lewis theory helps us to understand why solutions of   40
some metal ions are acidic. Cations hydrate in aqueous solution: this is essentially a Lewis acid–base reaction between the
oxygen of the water and the cation. This causes the withdrawal
of a lone pair from the O towards the cation, which makes the
O—H bond even more polar. When this effect is strong (as with   45
high charge-density cations), the O—H bond becomes so polar
that it can donate a proton to another water molecule, forming
H_3O^+. Thus an aqueous solution of chromium(III) nitrate has a
pH well below 7.

Adapted from Chemistry, The Central Science, *T. L. Brown and H. E. LeMay,*
Prentice Hall (1988)

1 Draw a Lewis structure (dot-and-cross diagram) of the product of
the reaction given in line 20. What special name do we give to the
N—B bond in this molecule?   [3]

2 Decide which of the following could act as a Lewis base, a Lewis
acid, or neither:
 a carbon monoxide;
 b the potassium ion;
 c the chloride ion;
 d argon;
 e hydrogen chloride,
briefly justifying your choice in all cases.   [10]

3 How does the Brønsted–Lowry theory define an acid? Find one
such acid in the list in question 2 and write an equation to
demonstrate its action as an acid.   [4]

4 Explain the meaning of the term 'vacant orbital' (line 18). [3]

5 What structural feature enables transition metal ions to act as Lewis acids? [2]

6 Could the addition of hydrogen bromide to ethene be regarded as an acid–base reaction in the Lewis sense? Justify your answer. [4]

7 Examine the equation in line 37, and use it to propose a mechanism for the reaction of water with sulphur trioxide to form sulphuric acid. [4]

Total = 30

Additional work

Read the last paragraph carefully, and then write a mechanistic equation to show why the hydrated chromium(III) ion is acidic.

Keto-enol tautomerism

experiment design, equilibrium, IR spectroscopy, mechanism, rates, structure

Electrophilic addition of water to ethene by a suitable method gives ethanol, and the reaction is an important industrial source of this compound:

$$CH_2\!\!=\!\!CH_2 + H_2O \rightarrow CH_3\!\!-\!\!CH_2OH$$

It might be expected, therefore, that reaction between   5
equimolar proportions of water and ethyne would yield ethenol. This is indeed the case, but the product is unstable and rearranges; equilibrium is established. The reaction, in which a hydrogen atom and a double bond effectively change places, is known as **keto-enol tautomerism**. It can be represented   10
mechanistically thus:

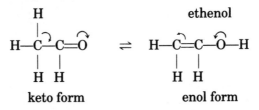

keto form   enol form

This equilibrium lies very far to the left. For practical purposes, there is no enol form present, and so addition of a further water molecule is prevented.   15

The phenomenon of reversible migration of a proton is possible in most carbonyl compounds, but in simple aldehydes and ketones the enol form is not normally detectable. Nevertheless, it is thought to play an important role in some reactions of carbonyl compounds: in propanone, for example,   20
only about one molecule in 10^6 is in the enol form, yet the reaction of this compound with iodine is believed to proceed via the non-carbonyl isomer:

$$CH_3CCH_3 \rightleftharpoons CH_2\!=\!CCH_3 + I_2 \rightarrow CH_2I\!-\!CCH_3 + HI$$
$$\underset{O}{\|} \qquad\qquad \underset{OH}{|} \qquad\qquad\qquad \underset{O}{\|}$$

The keto-enol rearrangement is known to be catalysed by H^+,   25
hence the enol mechanism proposed above is supported by the fact that iodine and propanone react very slowly in the absence of an acid.

In some more complex structures the enol form may be a significant proportion of the whole, up to around 90% in some   30
cases. The compound ethyl 3-keto butanoate, shown below in the keto form, contains about 7·5% of the enol form at room temperature.

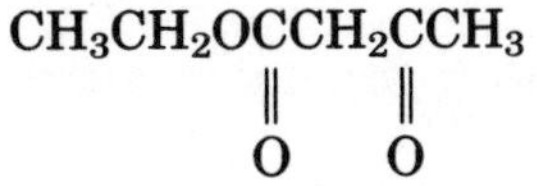

$$CH_3CH_2OCCH_2CCH_3$$
$$\underset{O}{\overset{\|}{}}\quad\underset{O}{\overset{\|}{}}$$

ethyl 3-keto butanoate

It is possible to measure the relative proportions of the two   35
forms chemically, e.g., by titration of the enol form with bromine under conditions in which the tautomerism reaction takes place very slowly (i.e., titration is complete before there is a significant shift in the equilibrium). However, it is more accurate and convenient to use infrared spectroscopy, since   40
carbonyl groups strongly absorb IR radiation, and the frequency of absorption will be slightly different for the two forms.

1 In the equation given on line 12,
   a give the systematic name of the keto form, and
   b explain the meaning of the 'curly arrows'.   [3]

2 Propose a structure for the hypothetical product of the addition of a further molecule of water (lines 14–15).   [2]

3 a Write an equation for the reaction of bromine with ethyl 3-keto-butanoate (lines 36–9).
   b If you were carrying out this titration, explain how you would cause the rearrangement to occur very slowly.   [4]

4 Explain why it is necessary to specify the temperature (lines 32–3).   [2]

5 Under what *conditions* is the addition of water to ethene (line 4) carried out on an industrial scale? Rewrite the equation mechanistically, and use it to show what is meant by 'electrophilic addition' (line 1).   [10]

6 In the equation given on line 24, of what *type* is the overall reaction? Explain clearly how the role of the proton (H^+) supports the proposed mechanism.   [5]

7 Draw a displayed formula of the enol form of this molecule (line 34). Compare the structure you have drawn with that of ethenol (line 12); can you suggest why the former is so much more stable than the latter? [4]

8 The author uses the word 'most' in line 17. Examine the equilibrium on line 12, and then draw the displayed formula of an aldehyde which could *not* exist in an enol form. [5]

Total = 35

Additional work

Describe the principle of IR spectroscopy, and show how it can be applied to measure the relative proportions of the isomers (lines 40–43).

Exercise 4·9

The dimerisation of ethanoic acid

*enthalpy, experimental technique, graph plotting and
interpretation, hydrogen bonding, ideal gas, K, safety*

At temperatures not far above its boiling point, the vapour of
ethanoic acid consists largely of *hydrogen-bonded dimers*, with
the degree of dimerisation decreasing as the temperature is
raised. In this experiment we employ Dumas' method to find
the equilibrium constant for the association, K_p, over a range of   5
temperatures, and from this standard enthalpy change of the
reaction, $\Delta H^{\ominus}$.

The equilibrium for n moles of ethanoic acid vapour associat-
ing to an extent α may be written:

$$CH_3COOH \rightleftharpoons \tfrac{1}{2}(CH_3COOH)_2$$ 10

Moles of vapour $= n(1 - \alpha)$ and $n\alpha/2$,

therefore total $= n(1 - \alpha/2)$

The experiment is carried out at the ambient atmospheric
pressure, p_{atm}. Assuming that the vapour is a perfect gas, we
may use the ideal gas equation: 15

$$p_{atm}V = n(1 - \alpha/2)RT$$

n is found by titration, and the volume, temperature, and p_{atm}
are also measured; thus α may be calculated.

Now, the partial pressures of the unassociated ethanoic acid,
p_1, and the dimer, p_2, are given by   20

$$p_1 = \frac{n(1 - \alpha)}{n(1 - \alpha/2)} \cdot p_{atm} = \frac{(1 - \alpha)}{(1 - \alpha/2)} \cdot p_{atm}$$

and $$p_2 = \frac{n\alpha/2}{n(1 - \alpha/2)} \cdot p_{atm} = \frac{\alpha/2}{(1 - \alpha/2)} \cdot p_{atm}$$

K_p can be calculated from these two. If K_p is found at a series of
temperatures, $\Delta H^{\ominus}$ for the reaction can also be calculated,
since   25

$$\ln K_p = -\frac{\Delta H^{\ominus}}{RT} + \text{const}$$

Method

Place some pure ethanoic acid (bp = 118 °C) in a small beaker, and draw about $2\,\text{cm}^3$ of it into a Dumas bulb (see the diagram), making use of a beaker of hot water. Measure the ambient pressure, p_{atm}.   30

A dumas bulb.

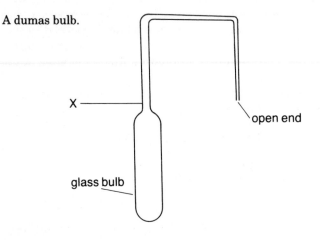

The following are results of a typical experiment.

T	= 557 K
p_{atm}	= $1.016 \times 10^5\,\text{N m}^{-2}$
	= 1.006 atm
Bulb volume V	= $5.38 \times 10^{-5}\,\text{m}^3$
Titre	= $52.6\,\text{cm}^3$
[Ba(OH)$_2$]	= $0.0119\,\text{mol dm}^{-3}$
(R	= $8.31\,\text{J K}^{-1}\,\text{mol}^{-1}$)

Place the bulb in a furnace at 150 °C, and collect the ethanoic acid in a test tube as it distils out of the open end. When all the liquid has distilled, and only vapour remains in the bulb, carefully wipe the open end to remove any liquid, and seal it by   35 melting the end with a blowtorch. Remove the bulb from the furnace and allow it to cool.

Place the end of the bulb under about $200\,\text{cm}^3$ water, and break open the seal; water will fill the bulb. Score the tube at X with a glass knife, and carefully break it off. Quantitatively   40 transfer the ethanoic acid solution to a conical flask, and titrate it against standard barium hydroxide solution. Finally, measure the volume V of the bulb by filling it with water from a burette.

The whole procedure is repeated at four or five more 45 temperatures up to around 250 °C, using a new Dumas bulb for each experiment. K_p is calculated for each temperature, and hence $\Delta H^{\ominus}$.

Adapted from Experimental Physical Chemistry,
G. P. Matthews, Oxford (1985)

Graph paper is needed for this exercise

1 Draw a displayed formula of the ethanoic acid dimer, indicating the hydrogen bonds. Why does the degree of dimerisation decrease as the temperature is raised? [4]

2 Describe in detail how you would transfer the ethanoic acid quantitatively to the conical flask (lines 40–41). [4]

3 What possible safety hazards are associated with this experiment, and what precautions would you take to minimise the risks? [3]

4 Write an equation for the titration reaction (line 42). What indicator would you choose, and why? [4]

5 Define an ideal gas (line 15). Is it reasonable to assume that ethanoic acid vapour *is* ideal under the conditions of this experiment? Justify your answer. [5]

6 Calculate values for α, p_1 and p_2 at 557 K. Write the expression for K_p, and calculate its value at this temperature. (Note: use p_{atm} in $N\,m^{-2}$ to calculate α, and in atm to calculate p_1 and p_2.) [11]

7 The values of K_p at four other temperatures are as follows:

$T(K)$	K_p	$T(K)$	K_p
428	0·8116	481	0·4622
450	0·6239	515	0·3425

Using the equation (line 26), and a suitable plot (include your own value for K_p), calculate a value for the enthalpy change of the dimerisation. What is the average bond energy of the hydrogen bond in the dimer? [14]

Total = 45

Catalysis and adsorption at surfaces

Arrhenius equation, enthalpy, ΔS, graph plotting and interpretation, rates

Almost the whole of the modern chemical industry depends upon the development, selection and application of catalysts, the majority of which are heterogeneous. The phenomenon of **adsorption**—the accumulation of particles at a surface—is crucial to such catalytic activity. 5

Molecules and atoms can stick to surfaces in two ways: in *physisorption* (an abbreviation of physical adsorption), there is a van der Waal's interaction (i.e., a dispersion or a dipolar interaction) between the substance which is adsorbed (the *adsorbate*) and the surface (the *substrate*). In *chemisorption*, 10 the particles stick to the surface by forming a chemical (usually covalent) bond, and tend to find sites which maximise their coordination number with the substrate. A chemisorped molecule may be torn apart at the demand of the unsatisfied valencies of the surface atoms, and the existence of molecular 15 fragments on the surface as a result of chemisorption is one reason why surfaces catalyse reactions.

The energy released when physisorption occurs is of the same order of magnitude as the enthalpy of condensation, whereas the enthalpy of chemisorption is very much greater, 20 and typical values are in the region of $-200\,\text{kJ mol}^{-1}$. Except in special cases, chemisorption must be exothermic, since the adsorbate loses translational freedom on adsorption, and hence ΔS for the system is negative: this must be outweighed by a corresponding positive value of ΔS for the surroundings. 25 The rate of chemisorption will depend upon the activation energy of the process, which is related to the increase in potential energy of the adsorbate as the bonds stretch prior to breaking.

Once adsorbed onto a surface, the particles must be able to 30 migrate if reaction with another adsorbate is to occur, so migration is a prerequisite of catalytic activity. Another essential for the substrate to act as a heterogeneous catalyst is that the adsorbed fragments should be more reactive than the original molecules. 35

After the catalysed reaction, the products must **desorb** from the surface. The activation energy, E_A, for this process can be determined by comparing the rates of desorption of the gas at a series of temperatures, T (K), and using the Arrhenius equation:

$$\ln k = C - \frac{E_A}{RT}$$

where C = a constant, k = the rate constant, and R is the gas constant.

One important example of heterogeneous catalysis is the hydrogenation of carbon-carbon double bonds, as in the production of margarine from vegetable oils. This is carried out for two reasons: first, double bonds are undesirable since they are subject to atmospheric oxidation, which causes rancidity, and second, hydrogenation of all or some of the double bonds raises the mp of the oil so that it is solid at room temperature—more convenient for making sandwiches!

Both hydrogen and the oil are adsorbed onto the surface, the former as H atoms, and the latter by using the π electrons ((1) in the diagram). A hydrogen atom migrates and replaces the C—substrate bond with a C—H bond (2). A second H atom then does the same with the second C—substrate bond, and the product desorbs (3).

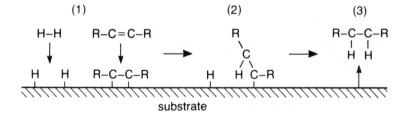

The mechanism of hydrogenation.

Many transition metals can catalyse this reaction, but the most active will be those in which the chemisorption is neither too weak (which means that little of the catalyst surface is covered with adsorbate) nor too strong. Ni, Pd, and Pt are the most effective catalysts, with Ni being the most widely used.

Adapted from Physical Chemistry, P. W. Atkins, Oxford (1986)

Graph paper is needed for this exercise.

1 Briefly explain the meaning of the following terms:
   a heterogeneous (line 3);
   b dipole interaction (lines 8–9);
   c coordination number (line 13);
   d π electrons (line 53). [8]

2 State and explain *one* other factor which will affect the rate of adsorption of a gas (lines 26–9). [3]

3 Suggest reasons why chemisorption which is too strong reduces the effectiveness of a catalyst (lines 59–62). [3]

4 What type of molecules are vegetable oils? Why should Ni, rather than Pd or Pt, be the most widely used catalyst for this process? [3]

5 What does the symbol ΔS (line 24) represent? Explain the precise connection between an exothermic reaction and a positive value of ΔS for the surroundings. [4]

6 What is meant by the *activation energy* (lines 26–7) of a reaction?
   The rate of desorption of a gas was measured at various temperatures, and the following data were obtained:

T (K)	k (s^{-1})
300	$1 \cdot 74 \times 10^{-6}$
310	$6 \cdot 61 \times 10^{-6}$
320	$2 \cdot 51 \times 10^{-5}$
330	$7 \cdot 59 \times 10^{-5}$
340	$2 \cdot 40 \times 10^{-4}$

Use the Arrhenius equation (line 41) and a suitable plot to calculate the activation energy of the desorption. $(R = 8 \cdot 31 \text{ J mol}^{-1} \text{ K}^{-1})$ [14]

Total = 35

Topic index

Answers to numerical problems

This page may be removed

2 Mainly about compounds

Ex.2·2 Q.2 molar ratio = 1:4, 18·8%
Ex.2·5 Q.2 $Z/r(Cs) = 0·006 \, \text{pm}^{-1}$
Ex.2.6 Q.1 $1·67 \times 10^{-17} \, \text{mol dm}^{-3}$

3 Organic chemistry

Ex.3·2 Q.3 65%
Ex.3·3 Q.6 5·6 g
Ex.3·4 Q.2 26%
Ex.3·7 Q.1 17·4
Ex.3·8 Q.2 24% C, 76% F
        Q.7b $-114 \, \text{kJ mol}^{-1}$

4 Energy, equilibrium and kinetics

Ex.4·3 Q.4 $-3218 \, \text{kJ mol}^{-1}$
        Q.6 fuel value = $15·18 \, \text{kJ g}^{-1}$
             $\Delta H^{\ominus} = -1366 \, \text{kJ mol}^{-1}$
Ex.4·4 Q.4 $5·4 \, \text{mol dm}^{-3}$
Ex.4·6 Q.4 $-187·8 \, \text{kJ mol}^{-1}$, $-37·6 \, \text{kJ mol}^{-1}$
        Q.7 $\pm 0·14 \, \text{V}$ (depending on cell diagram)
Ex.4·9 Q.6 $\alpha = 0·1135$
             $p_1 = 0·945 \, \text{atm}$, $p_2 = 0·061 \, \text{atm}$
             $K_p = 0·2614$
        Q.7 $\Delta H_{\text{dimerisation}} = 16·0 \, \text{kJ mol}^{-1}$
Ex.4·10 Q.6 $107 \, \text{kJ mol}^{-1}$